MR SMITH
PROPAGATES PLANTS

by Geoffrey Smith

Illustrated by Colin Gray
Edited by Brian Davies

BRITISH BROADCASTING CORPORATION

Published to accompany the BCC-tv series *Mr Smith Propagates Plants* produced by Brian Davies and Peter Riding and first broadcast on BBC-2 from March 1978

Published to accompany a series of programmes prepared in consultation with the BBC Further Education Advisory Council

Other BBC books by the author

Mr Smith's Gardening Book
Mr Smith's Vegetable Garden
Mr Smith's Flower Garden
Mr Smith's Fruit Garden
Mr Smith's Indoor Garden

The photographs on front and back covers and page 4 were specially taken by Jon Blau, Camera Press

© Geoffrey Smith and the British Broadcasting Corporation 1978
First published 1978. Reprinted 1978, 1979 (twice), 1980
Published by the British Broadcasting Corporation, 35 Marylebone High Street, London W1M 4AA
Set in 9/10pt Plantin
Printed in England by Alan Pooley Printing Limited, Tunbridge Wells, Kent
ISBN: 0 563 16210 4

Contents

Introduction   4
Preface, tools   5

SOWING SEEDS

Compost   6
Soil based and peat based compost   7
Containers   8
Sowing indoors   9
Sowing in a frame   10
Sowing outdoors and sowing vegetable seed   11
Pricking out and potting off   11
Rules for sowing   12
Stratification of seeds   12
Alphabetical list   14-19

TAKING CUTTINGS

General principles   20
Stem or top, and soft wood cuttings   21
Semi-hardwood cuttings   22
Hardwood cuttings   23
Outdoor and garden frame for cuttings   24
Rules for taking stem cuttings   25
Leaf cuttings   25
Leaf bud and root cuttings   26
Scales   27
Plants with special needs   27
Pipings   28
Alphabetical list   30-33

LAYERING

Branch layering   34
Tip layering and runners   35
Serpentine layering   36
Air layering and stools   37
Alphabetical list   38

DIVISION

Bulbs, corms and suckers   40
Plants with special needs   41
Alphabetical list   42-43

BUDDING AND GRAFTING

Budding   44
Chip budding   48
Whip, tongue and saddle grafting   49
Rind or Crown grafting   50
Bridge grafting   50
Alphabetical list   51

Index   52-56

Introduction

One of my earliest recollections of seed sowing is concerned with growing beans in a 2lb jam jar, lined with blotting paper on a village school window sill. Since then I have sown many different types of seed and taken thousands of cuttings, but the fascination remains.

To see an apparently dead husk push up a green shoot, or watch a piece of twig casually pushed into compost develop new roots and grow new leaves is still amazing.

From the busy season of late winter, early spring through the rush of taking stem cuttings in summer, to the more leisurely business of selecting hardwood and root cuttings in autumn, the propagation of plants offers a perennial and absorbing interest.

Geoffrey Smith

Preface

Once a gardener, professional or amateur has a plant in his care the first question may well be 'What sort of growing conditions will it need?'. The second is usually 'How can it be propagated?'.

A lot of the work done in the garden is seasonal, mowing the lawn and pruning the roses for example, but plant propagation goes on the whole year round, particularly if the interest includes house plants.

There is no aspect of gardening more interesting than that which involves raising new stocks of plants from seed or by means of cuttings, grafting or budding.

Seed is the sexual method, the universal means by which flowering plants ensure their survival and increase their number. Sowing seeds offers the gardener an easy, rapid means of acquiring a large stock of a particular species. Only species, that is a group of plants showing the same characteristics, can be guaranteed to provide seedlings identical with the parents. For example, seed sown from a 'Dog Rose' (Rosa canina) will, in due course, grow into plants identical with the parents. Seed saved from a hybrid rose, for example, 'Evelyn Fison' will not breed true to the parent but grow into plants which may vary in height, leaf and flower colour. So to increase ones stock of hybrid plants, the gardener must resort to budding, grafting, layering, division or cuttings, all of which will be described in detail later in the book.

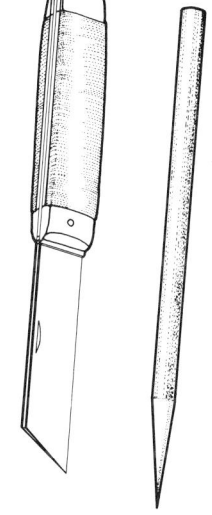

Tools

To propagate plants successfully, you do need certain basic tools and I find the collection of useful gadgets tends to increase in proportion to the interest in extending one's stock of plants. To begin with a sharp knife is essential: so far I have found no substitute. A smooth, round blunt ended piece of wood or plastic ½-¾ inch in diameter to use as a dibber is useful. On occasions, when nothing else offers I have used a pencil or my finger to make holes when pricking out seedlings. I now have a selection of tools which are invaluable aids to seed sowing, taking cuttings and grafting: old dinner forks for lifting seedlings, a desert spoon for potting, eyebrow tweezers for spacing seeds, and a dibber which started life as a pricker for making wool mats.

When preparing composts for seed sowing a riddle for sifting the material is a great help. A good general purpose riddle would be one which graded to a ⅜ or ½ inch grist. For really fine sowing I make a riddle out of four pieces of wood, 12 inches long by 4 inches broad by ¼-½ inch thick . Nail these together to form a square then instead of a solid base, tack on a piece of perforated zinc.

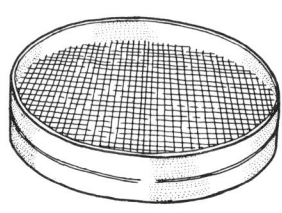

Sowing Seeds

Raising plants from seed requires no specialised knowledge and very little equipment — after all plants do it naturally. Seeds of most plants, in addition to the embryo, contain enough food to support the seedling until it has developed sufficiently to survive independently. All the gardener has to do is provide the sort of conditions in which germination and growth can take place to maximum effect. To germinate, most seeds need moisture, oxygen, a sufficiently high temperature and, eventually, a readily available food supply.

As a general rule, seeds retain full viability and vitality for a limited period only, so are best sown within twelve months of harvesting unless stored under special conditions. Modern techniques of refrigeration and air sealing can extend the storage time indefinitely. Frequently, the gardener will need to harvest and store seed from some favourite plant which is not in general commerce. When gathering seeds from your own garden, make sure they are thoroughly ripe. Before storing, spread them out to dry on a tray in a shed or similar dry, airy place. Clean the seed after drying off all pods, husks, or similar unnecessary debris. Then store in a cool, dark place where the temperature is even. Shiny paper envelopes make good containers and the name of the seed can be written on the outside.

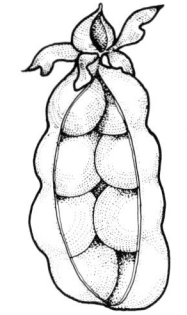

1

Seeds differ widely in shape, size and texture. Some such as begonia are dust-like, while others are large enough to be handled individually. Oily seeds such as magnolia, walnut, paeony (1), chestnut and oak are best sown immediately they ripen. If stored in the usual way they tend to shrivel, but this can be prevented by burying them in moist sand. This must be for a limited period only or they start to grow. Japanese quinces (2) can be stored complete until the springtime.

Seeds of most shrubs and hardy plants are best stored in a cool, dark place until the following spring. They can be sown under glass in February or March, or a month later if sowing outdoors.

Buying your seeds from a reputable nurseryman is a good way of ensuring they will grow well and be true to name.

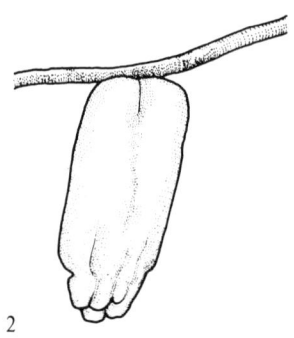

2

COMPOST

Because the various ingredients required for making up seed composts are not easy to obtain in small quantities you may find it easier to buy the composts ready mixed. Broadly speaking, there are three types to choose from. One is based on soil, another on peat. They are both suitable for growing non-specialist plants — annuals, vegetables etc. The third is a lime free or ericaceous mixture specially prepared for rhododendrons, gentians and similar more specialised garden plants which die if there is lime present in the compost.

All composts, no matter what their individual ingredients are, should be free from pests, diseases, and weeds. They should hold sufficient moisture for the plants' needs without becoming stagnant. Finally, they should contain a steady supply of food to support the seedlings in balanced growth.

Soil based compost

One of the problems facing the self-sufficient gardener who wishes to mix his own soil-based compost is where to buy good quality loam. For those fortunate enough to locate a source of turf loam — turf loam is made of turf, cut from old pasture land and stacked for 18 months before use (1) — a useful *seed* compost could be made up as follows, all parts by volume except fertilizer which is by weight. 2 parts loam passed through a three-eighths inch sieve. The loam should be sterilised either with chemicals or by being heated to a temperature of 180 degrees F., 82 degrees C. (2) (3) and kept there for 10 minutes. After sterilising, tip the loam onto a *clean* surface to cool. The remaining ingredients are 1 part moss peat of medium horticultural grade; 1 part lime free sand fairly coarse, up to ⅛ inch grist. To each bushel (this is a measure 22ins. x 10 ins. x 10ins. filled level without firming) add 1½oz. superphosphate and ¾oz. chalk.

For plants which dislike lime, azaleas for example, omit the chalk, or preferably mix a special compost based on peat. Otherwise the compost will suit almost all plants, shrubs, bulbs, herbaceous, annuals and vegetables.

Peat based compost

Peat or peat and sand composts were introduced because good quality loam is hard to find in quantity. Seedlings grow quickly in a well made peat compost, but care is necessary to ensure they never lose too much moisture, for once dry it is difficult to wet them again except by soaking the pots or boxes in a tub of water. Conversely, with a peat compost, avoid overwatering. Peat composts, especially if they are tightly compacted, do waterlog, particularly in the winter.

There is very little to be gained in trying to economise on the materials used for making compost. To achieve consistently good germination and growth of seed, the loam, peat and sand must be of top quality. They must be mixed in the right proportions and stored carefully away from possible re-infection from harmful pests or disease. Heavy duty polythene bags make good containers for storing compost in.

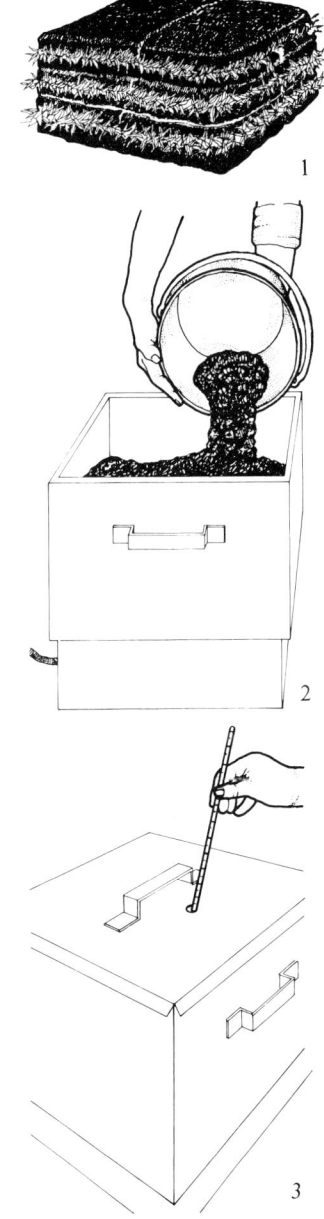

1

2

3

When making your own compost, do make sure all the ingredients are thoroughly mixed together before use. Spread the loam out on a clean, level surface, then the peat, finally the sand (1) and sprinkle the fertilizer evenly over the heap (2). Turn the heap over three times to mix the materials (3). Peat based composts are available ready mixed and the quality is fairly constant. Unfortunately the same cannot be said for loam based composts the quality of which fluctuates considerably.

Compost should be used within two months of buying or preparing it, otherwise the chemical character may be changed as the fertilizer breaks down.

One of the most common mistakes made when preparing a compost is to add extra fertilizer on the assumption that the richer mixture will produce a bigger plant in a shorter time. Frequently the excess fertilizer just poisons the young germinating seed. Should the seedling survive, the extra food causes the sort of lush, soft, growth which is readily prone to attack by disease. The most important period in a plant's life are the weeks immediately after germination. During this stage there must be no check to growth, especially from under or over nourishment. No care in regard to feeding in afterlife can compensate for a setback at this stage.

Phosphate is particularly important in the period immediately after germination as one of the things it stimulates is healthy root growth. Calcium neutralises excess acidity and helps the superphosphate work more efficiently. Before the seedlings need extra food, they will be large enough to transplant into John Innes potting compost which contains larger amounts of nitrogen, phosphates and potash.

The standard peat based composts bought ready mixed are advertised as being suitable for seed sowing or potting. Supplementary feeding may be necessary with seedlings of sweet peas, paeonia or shrubs which are grown on several weeks before potting off. Because there is no loam in the compost, fertilizer is soon washed out with watering.

CONTAINERS

Any receptacle which is clean and will hold the moist compost without collapsing will serve as a container for the seed bed. Pots, either clay or plastic, trays made from wood or plastic, trifle containers or any of the wide variety of vessels used for pre-packed food, will be suitable. It is better to choose one type so that all the seedlings can be watered as a whole. Where plastic, wood and clay containers are mixed up together problems arise because each will need different amounts of water. By settling on a standard container not only will watering be simplified but the whole layout will look neater.

SOWING INDOORS

Pots or boxes specially designed for seed will have proper drainage holes in them. Excess moisture must be able to drain away quickly and easily or the compost becomes waterlogged; a condition which can kill the seedling. No matter what container is used it must be clean and sterile.

Cover the drainage hole at the bottom of the container with a piece of broken clay pot (crock), a handful of gravel or, if available a piece of perforated zinc. This prevents the compost washing down to block the drainage and stops worms getting in to disturb the seedlings. A layer of rough peat on top of the drainage material will give an additional insurance against the fine compost washing down.

Fill the pot with seed compost, firming lightly with the fingers. Do not over compact-this is particularly important with peat based composts. The finished level should be a half or three quarters of an inch below the pot rim, depending on the type of seed being sown. A final exact level is made using any object with a smooth even surface, such as a milk bottle, tin or smooth piece of wood (1).

When sowing very fine seeds this levelling is important as an uneven surface causes some of the seed to be sown too deep. The absolute beginner would be well advised to sieve a little fine sand as a final cover through a perforated zinc riddle (2).

All seeds must be sown thinly (3). Overcrowded seedlings, starved for food, light and air, fall prey to disease. Even with practice, it is hard to sow fine seed like lobelia or begonia evenly, so try mixing it with fine sand or crushed brick dust. Fine seed is best just firmed lightly into the surface of the compost with no further covering.

The length of time a species or variety takes to germinate depends not only on temperature but the type of seed and time of year. Some annual and vegetable seeds show growth in just a few days. Others, such as holly or roses, take up to twelve months to appear, unless treated with chemicals or given a few weeks exposed to below freezing temperatures.

No seed should be covered with more than its own depth of soil or all the food supply contained within the embryo is used up before it breaks through the surface of the compost. Large seeds such as sweet peas can be handled individually — spaced out over the surface of the compost and then covered with their own depth of compost (4).

Many seedsmen offer pelleted seed. Such a seed has a coating of water-soluble material around it with fertilizer added. This increases their size so they can be spaced accurately in the compost. Pelleted seed must be kept moist after sowing or the hard coat will not break down. This could prevent growth. After sowing, cover the seed either with sharp sand or the riddled compost (1), then water. This first watering must ensure the compost is thoroughly and evenly moist. Immerse the container up to the rim in a bowl of water (2) until the moisture oozes out of the surface. It is a good idea to water the compost BEFORE sowing fine seeds which are not covered as this prevents them all washing to the side of the seed container.

Label the container clearly, showing the name of the seed and the date of sowing. Cover the seed tray or box with a sheet of newspaper then, if available, a pane of glass. Usually no further water need be given until the seeds have germinated.

Once the seedlings appear remove the glass and paper to prevent them becoming drawn by lack of light. During warm weather, spray over in the morning and again at evening for at no time must they be short of water. DO NOT discard a seed box if there is no sign of growth after a few weeks — be patient!

All seeds need a warm, even temperature to grow well. In general, 60 degrees F. will be adequate. There are some tropical plants whose seed will not grow at temperatures below 65-70 degrees F. An airing cupboard will provide a suitable nursery, but immediately the shoots show above the compost they will need moving out into daylight.

SOWING IN A FRAME

When there is no greenhouse available, seed may be sown in a frame (3). Unless this is heated, delay sowing the seed until the weather starts to warm up in late March. Otherwise the technique is identical with that used for greenhouse propagation.

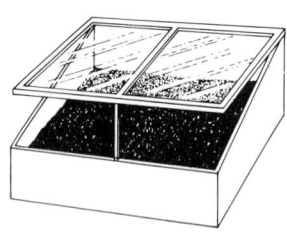

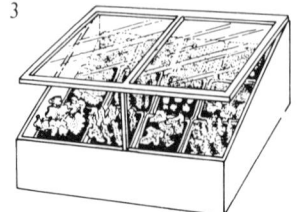

Try to site the frame in a sunny position but sheltered from strong winds which can make ventilation difficult. When the frame is bedded onto soil, the seed may be sown direct into the soil. Early salad crops, cabbage, cauliflowers, leeks and other vegetables may all be started into growth this way.

Seeds of certain species, which can take months to germinate, are less trouble when grown in a frame, even if it is outdoors, provided it is protected by a hedge or building. Germination of seeds such as gentians, meconopsis, primula, roses and cotoneaster can be hurried along if they are first sown then exposed to frost or given two or three weeks in the refrigerator. The old fashioned method of stratifying seeds like holly, hawthorn, cotoneaster and berberis between layers of sand outdoors is still a practical solution. Protect the pots by covering with netting (1) because mice will eat any seeds they can find.

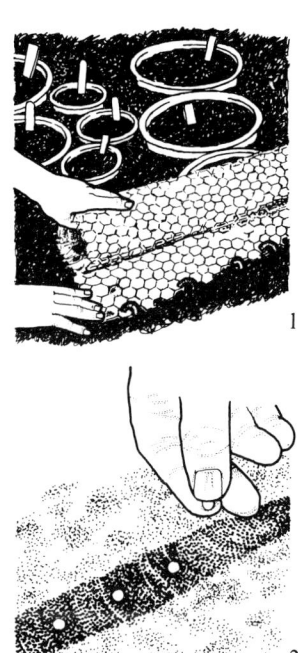

SOWING OUTDOORS

Sowing seed in the open ground is nature's way, indeed, the only way if the garden is without frame or greenhouse. There is, of course, less control over growing conditions so it is only sensible to wait until the weather is warm, in May or even June in colder areas. The soil needs to be warm and moist, but not so wet that it cakes down hard when the sun gets to it.

Choose a sheltered border which receives a fair share of sun. If the soil is at all heavy, work in a generous dressing of peat and sharp sand plus a dusting of superphosphate of lime at the rate of 1oz per sq. yd. This will help to build up a strong root system. Then, rake and cultivate the area until the soil is in a fine, friable condition. Sow the seed in shallow drills (2) rather than scattering it evenly over the whole bed. Having the seed in evenly spaced straight lines makes weed control much easier. As with indoor sowing, do not bury the seed too deep. Cover them by raking lightly along the row, then firm the soil by tamping down the length of the row with the back of the rake (3). One of the advantages about open ground sowing of shrubs or perennials is that they can be allowed to grow on undisturbed for at least a year after germination, provided they are not overcrowded.

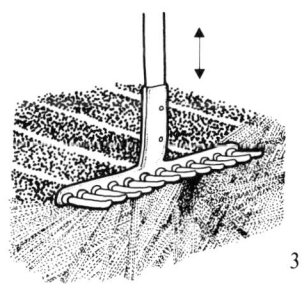

Label each row clearly and keep a plan of the nursery border in case some of the labels are lost. Water the seed regularly in hot dry weather and use the hoe to control weeds.

SOWING VEGETABLE SEED

Probably the most common form of open ground sowing is concerned with vegetable growing, most of which are grown from seed and treated as annuals. Rough dig the soil in autumn, working in compost or manure. Then, after being exposed to the action of the frost all winter, it should rake down into a good tilth for seed sowing. Early in the spring crops can be achieved by using cloches to protect the seedlings, but apart from this the technique of sowing is the same for vegetables as for ornamental plants.

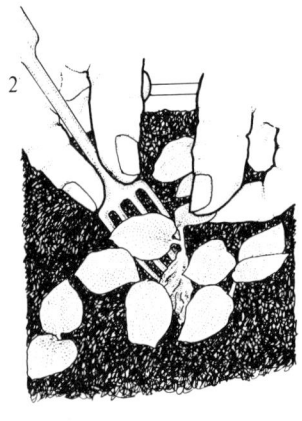

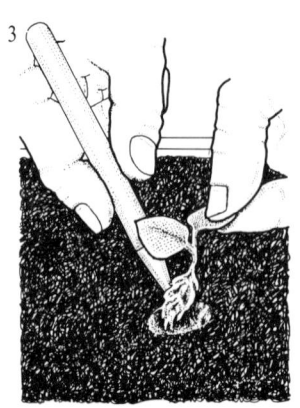

PRICKING OUT AND POTTING OFF

Pricking out and potting off are terms used for the transferring of seedlings from the container in which they were sown into the one in which they will grow on, possibly to maturity. There is only a limited supply of food in the seed compost and, except in exceptional circumstances it is good garden practice to transfer the seedlings when they are large enough to handle into a larger container filled with richer, potting compost.

For this purpose a good soil based compost would be: 7 parts, by bulk, of loam passed through a ⅜ ins. sieve, after being sterilised as described on page 7 for seed composts, 3 parts, by bulk, of horticultural medium grade peat, and 2 parts by bulk of coarse sand, up to ⅛ ins. grist.

To each bushel of the mixture add a quarter pound of the John Innes Base fertilizer, which is available under this name, and ¾ oz of chalk or ground limestone. OMIT the ground limestone or chalk if the plants being pricked off dislike an alkaline soil.

An old table fork makes a useful tool for lifting the seedlings. Insert it into the compost far enough away from the seedling so as not to damage the stem (1). Then, lever gently so the root system comes up more or less intact (2). Next, make a hole large enough to receive the root in the stronger compost (3). Ease the roots gently in and firm the soil back, right to the base of the roots so that no air pockets are left. NEVER handle seedlings by their stems, always by a leaf tip.

Begonia or lobelia seedlings are so small any attempt to separate them so they can be pricked out individually would damage them. They are pricked off in small groups of 3 or 5.

After pricking out or potting off, keep the seedlings in a close atmosphere well watered and shaded from the sun until the roots have penetrated into the new compost. Make sure the plants are all clearly labelled with their names and the date of sowing. This will help to avoid confusion at planting time.

RULES TO FOLLOW WHEN SOWING SEED

Choose a suitable compost: one free from disease, pests and weed seed and with a supply of food to keep the seedlings in balanced healthy growth.

Keep them warm, correctly watered and prick them out into a richer compost in good time.

Use only healthy seed from a reputable source.

STRATIFICATION

Some seeds have such a hard protective covering that they need special treatment to break this down to enable germination to take place. This process is called stratification.

There are other seeds which unless sown immediately they are ripe, although they have no hard outer covering, also need stratification to break dormancy.

The hard coated seeds fall into two classes. Those like holly, cotoneaster, peach stones etc. are buried in sand-filled pots or trays (1) which are then placed in an out-of-the-way corner for twelve months. The alternate freezing and thawing work on the hard coating so that, when the seeds are sown twelve to eighteen months later, germination takes place. Some hard coated seeds only need carding or chipping with a sharp knife, piece of sandpaper or stiff wire brush. Sweet peas, paeony, and magnolia can all be helped in this way.

Rose seed needs stratification before it will germinate, a fact which is frequently overlooked. Seed from hybrid roses will not come true to the parent but there is the possibility that one of the seedlings will be outstandingly good. A brief period in a freezing temperature or stratification in sand will break dormancy in this case.

Some seed which unless sown immediately they are ripe go determinedly dormant. These include gentians, primula, meconopsis, cyclamen and delphinium. To make old seed grow, expose them to below freezing temperatures by sowing them (2) and standing the pot outdoors during winter (3). Alternatively, they can be shaken up in moist sand then put in the refrigerator for a week or two.

All sorts of techniques are worth trying in order to encourage hard coated seeds to germinate. A process which I have adopted with good effect, particularly with hard coated seeds from South Africa and Australia, is to burn them in. Sow the seed on top of a prepared compost in the usual way. Put a handful of DRY hay or bracken on top and then set fire to it. The seed is then covered with compost and watered in the usual manner. Hay or bracken must be bone dry so that it produces a quick "flash" fire or the seed is killed.

Lime-hating plants should only be grown in an acid compost. A mixture of 2 parts finely sieved peat to 1 part of sharp lime-free sand is the one I use. Make sure the drainage is good by putting gravel or pieces of 'crock' in the bottom of the pot, then fill with the compost. After firming, the level should be approximately ¼ ins. below the pot rim. Stand the pot up to the rim in a bowl of water until the compost is evenly wet all through before sprinkling the seed thinly across the surface.

Lightly firm but do not cover except with a piece of newspaper and glass which is removed after germination. Shade the young seedlings from direct sunlight in the early stages of growth. This is the method I use for rhododendrons, pieris, calluna and erica.

ALPHABETICAL LIST

To give anything but the most abbreviated list of plants which can be raised from seed would be impossible. All plants which produce seed can, of course, be propagated by this means. All the seeds listed below should be sown into a standard seed compost unless stated otherwise. The term greenhouse should be taken to mean any light dry place which can be kept at a constant temperature.

Ageratum Sow in February-March in a heated greenhouse.

Alyssum Sow in February-March in a heated greenhouse.

Amaranthus Sow in February-March in a heated greenhouse.

Antirrhinum Sow in February in a heated greenhouse.

Aster Sow in March in a heated greenhouse in seed compost or into individual peat pots.

Aubrieta Sow in May into cold frames or open ground in seed compost.

Azalea Sow when ripe or in March onto the surface of a lime free peat compost. DO NOT cover the seed except with newspaper or glass.

Begonia Sow in January-February onto the surface of a standard seed compost. DO NOT cover the seed except with newspaper or glass.

Berberis (species) Seed sown outdoors into nursery beds when ripe. Protect seed from mice until they have germinated.

Cactus (species) Sow in a heated greenhouse in January-April in sandy compost.

Calceolaria Sow as pot plants under glass in May-July. For summer bedding sow in February-March in standard seed compost.

Calendula Sow outdoors direct where they are to flower in spring-summer.

Campanula Sow, depending on the type, in spring-early summer. Just press seed into the soil surface without covering.

Cheiranthus (Siberian Wallflower) Sow in May-June in open ground or in frames.

Chrysanthemum (annual) Sow in January-March in a heated greenhouse.

Cineraria Sow in December to April in heat. Then May to August to give a succession of blooms. DO NOT bury seed too deeply.

Clarkia Sow in April-May direct where they are to flower.

Clary Sow in April-May direct where they are to flower.

Coleus Sow in January-June in a heated greenhouse.

Convolvulus Sow in April-May where they are to flower.

Coreopsis Sow in April-June into the open ground.

Cornflower (Centaurea) Sow April-May direct into the open ground where they are to flower.

Cosmos Sow in April-May outdoors where they are to flower.

Cotoneaster Stratify seeds then sow outdoors in nursery beds in March-April.

Crataegus (Hawthorn) Stratify the berries. Then sow outdoors in nursery beds in spring some eighteen months later.

Cyclamen Sow in February-April in a heated greenhouse in standard seed or potting compost.

Daphne Gather and sow the seeds when ripe into a cold frame in standard compost. They may take a year to germinate.

Delphinium Sow the seed into nursery beds outdoors in May-June.

Dianthus (Carnations & Pinks) Sow in January-April under glass in a standard seed compost.

Digitalis (Foxglove) Sow the seed outdoors in spring.

Eschscholtzia Sow in April-May, outdoors, where they are to flower.

Euonymus Sow the seed in March in a cold frame.

Freesia Sow in a heated greenhouse March to April, or in a cold frame May-June.

Gentiana Seed sown when ripe in a lime free compost. Stand in a cold frame. Exposure to frost improves germination.

Geranium (species) Seed sown in March-April. Stand in a cold frame.

Gloxinia Sow in January-March in a heated greenhouse.

Godetia Sow in April direct where they are to flower.

Grasses (ornamental) Sow direct into ground April-May.

Gypsophila Sow in April-May in a frame.

Hamamelis Seeds sown when ripe into a cold frame. They may take 18-24 months to germinate.

Helleborus Sow seeds when ripe or in the spring into pots standing in a cold frame. Use a sandy compost.

Hollyhocks (Althea) Sow in April-May in a seed bed outdoors.

Honesty (Lunaria) Sow in May-June direct where they are to flower.

Iberis (Candytuft annual). Seed sown direct where they are to flower in the open ground.

Impatiens Sow in March in a heated greenhouse.

Incarvillea Sow in March-April into a frame or sheltered border outdoors.

Ipomoea (Morning Glory) Sow in March after soaking seeds overnight in water in a greenhouse or frame.

Laburnum Seeds sown into a prepared bed outdoors when ripe.

Larkspur Sow in April-May direct into the open ground where they are to flower.

Lavatera Sow in April-May direct into the open ground where they are to flower.

Lilies Sow the seed into a cold frame, spring to early summer.

Limnanthes (Foam of the Meadow) Sow in April-May direct where they are to flower.

Linum Sow in April-May direct where they are to flower.

Lobelia Sow in heat January-March.

Love Lies Bleeding (Amaranthus) Sow in April-May direct where they are to flower.

Lupin Sow the seeds into a cold frame made up with sandy compost in March or outdoors April-May.

Lychnis Sow in April-June in the open into a prepared seed bed.

Magnolia Because they are oily, the seeds soon deteriorate. Sow in pans filled with standard seed compost immediately the seed is ripe.

Marigold (African & French) Sow seed in heat, March-April, or in the open during May.

Matricaria Sow in March in heat, standard compost.

Matthiola (Night Scented Stock) Sow in April-May direct where they are to flower in open ground.

Meconopsis Sow immediately the seed is ripe into pans of lime free compost and stand in a cold frame. Old seed should be sown and the pans exposed to frost.

Mesembryanthemum (Livingstone Daisy) Sow in February-March in heat in standard seed compost.

Mignonette (reseda) Sow in April-May where they are to flower in the open ground.

Mimulus (Monkey Musk) Sow in February-March in a heated greenhouse.

Myosotis (Forget-me-Not) Sow in May-June in the open garden to flower the following spring.

Nasturtium Sow in April-May direct where they are to flower in the open garden. They thrive in poor, dry conditions.

Nemesia Sow in February-April in a heated greenhouse.

Nicotiana (Tobacco Flower) Sow in February-April in heated greenhouse.

Nigella (Love in a Mist) Sow in April-May direct where they are to flower in the open garden.

Oenothera (Evening Primrose) Sow in a heated greenhouse during March or outdoors in June.

Paeony Sow the seeds of species when ripe into a cold frame.

Pansy Sow in a heated greenhouse during February-March in standard seed compost. Alternatively, sow in a frame during June then they flower the following year.

Papaver (Poppy) Sow the annual varieties, in April-May direct where they are to flower.

Penstemon Sow in March-April in cold frame.

Petunia Sow in February in heated greenhouse.

Phlox (annual) Sow in February-April in heated greenhouse.

Polyanthus Sow in heat during February, or in a cold frame in April.

Primrose Sow in April-May in a frame, standard seed compost.

Primula (greenhouse) Sow in May in greenhouse. Use a peat based compost.

Primula (hardy) Sow seed immediately it is ripe into a cold frame in peat based compost. Old seed should be exposed to frost or it may take a year to germinate.

Pyrethrum Sow in May-June in a cold frame, standard compost.

Rose Seed of rose species will germinate after twelve months if sown immediately it is ripe into standard compost. May also be stratified in sand.

Rudbeckia Sow seed in February in heated greenhouse, standard compost, or direct where they are to flower in June.

Saintpaulia (African Violet) Sow seeds in a heated greenhouse in late winter. Use a peat based compost — a warm humid atmosphere quickens growth.

Salpiglossis Sow in February-March in a heated greenhouse. Outdoors, in May direct where they are to flower.

Salvia Sow singly into small peat pots during February-March in a heated greenhouse.

Saxifraga Sow in March into pans in a cold frame, use a standard seed compost with extra sand. 2 loam, 2 sand, 1 peat.

Schizanthus (Butterfly Flower) Sow in August for spring flowering or March for summer bedding.

Solanum capsicastrum (Winter Cherry) Sow in February in heated greenhouse.

Statice Sow in February-March in a heated greenhouse.

Stocks (Ten-week). Sow February-March in heat, or direct where they are to flower in April.

Stock—East Lothian Sow in March in heated greenhouse for summer bedding, or in July to flower under glass as pot plants the following spring.

Sweet Pea (Lathyrus) Sow in January in a heated greenhouse; April in the open ground where they are to flower; or September to overwinter in a cold frame. Use a John Innes No.1 compost. A strong root system is particularly important when growing Sweet Peas for early cut flowers so it is common practice to use a longer pot. Special tubes can be bought for Sweet Peas but I find the long narrow disposable plastic or cardboard cups do just as well. Bore a hole in the bottom for drainage before sowing the seed.

Sweet William Sow seeds in June outdoors to flower the following year.

Tagetes Sow February-March in heat. For a late display of flowers seed may be sown direct where the plants are to flower outdoors from mid-May onwards.

Trollius (Globe Flower) Sow in March into peat based compost in a cold frame.

Verbena Sow in February-March indoors.

Viola Sow in February-March in heat. Sow in the open ground during July for spring flowering.

Wallflower Sow in May-June in the open ground in a well limed soil. Transplanting the seedlings from the nursery to a lining-out bed for growing on is good practical gardening. This breaks the tap root and encourages the formation of a strong fibrous root system. Space the young plants 3-4 inches apart in rows across the vegetable plot after the early potatoes are lifted.

Zinnia Sow in March in heat using standard compost. Or

May into the open ground where they are to flower. Zinnia resent root disturbance, so if possible grow the seedling singly in small pots. The soil blocks or peat pots are very suitable.

Vegetables

Most of the vegetables commonly grown in gardens are, of course, grown from seed. Nearly all are sown direct into prepared soil in spring, except tender crops such as tomatoes, melons and peppers. Prepare the soil prior to sowing by raking in a light fertilizer dressing and wait till the soil has warmed sufficiently in spring. All the following are hardy and may be sown outdoors.

Beans	Broad—Rich soil. February onwards.
	French—Rich soil. May.
	Runner—Rich soil. May.
Beetroot	Firm soil, no fresh manure. May onwards.
Broccoli	Firm soil, no fresh manure. May onwards.
Brussels sprouts	Firm soil, no fresh manure. March onwards.
Cabbage	Rich soil. March onwards.
Carrots	Sandy soil. February onwards.
Cauliflower	Rich soil. April onwards.
Chicory	Rich soil. May onwards.
Kale	Firm but not too rich soil. April.
Leeks	Rich soil. Sow April.
Lettuce	Rich soil. Sow April.
Onions	Rich soil. March.
Parsnips	Moderately rich soil. March.
Peas	Rich soil. Sow February onwards.
Radish	Rich soil. Sow February onwards.
Spinach	Rich soil. Sow February onwards.
Swedes	Well limed soil. Sow April-May.
Turnip	Well limed soil. Sow April-July.

The following are not hardy and are better sown under glass then transferred to the open ground when all fear of frost has gone.

Aubergine	Sow April—under glass.
Celery	Sow March—under glass.
Courgettes & *Marrows*	Sow late April—under glass.
Cucumbers	Sow February onwards —under glass.
Peppers	Sow April—under glass.
Tomatoes	Sow January onwards —under glass.

Taking Cuttings

The word cutting is a collective term used to describe any portion of plant which, after being severed from the parent can be persuaded by careful cultivation to grow into a completely new individual. A cutting can be any portion of a root, stem, or leaf. It differs from layering or division which already have a fully formed root system before being separated, because the cutting is initially devoid of roots.

There are several different kinds of cuttings, each needing particular treatment and cultivation according to which part of the plant it is taken from. Gardeners, for convenience sake, class cuttings under headings.

Stem or **Top cuttings** (1) can be taken from any non-flowering side shoot which because they are available almost at any season and stage of growth are sub-divided, for easier identifiction into softwood, semi-hardwood and hardwood.

Leaf cuttings (2) are chiefly used to propagate certain greenhouse plants such as begonia, gloxinia and streptocarpus, but can be tried with any perennial with fleshy leaves.

Leaf-bud cuttings (3) are a cross between a leaf and stem cutting in that it includes a leaf with a bud at the base and a section of stem.

Root cuttings (4) are used mostly to increase stocks of plants which have fairly thick roots and are, as a general rule, taken when the plant is dormant. Not all plants with thick roots can be increased in this way and some plants with thin roots can be propagated by means of root cuttings. So it is impossible to generalise.

Scales are a method of propagation used to increase stocks of some lily species.

General principles which apply to all cutting taking
Use only material from strong healthy plants which are true to species or variety.

Do not leave cutting material to dry out before inserting them in rooting compost. If it is impossible to avoid delay stand the cuttings in water or wrap the ends of the shoots in moist paper or sphagnum moss, then put them into a polythene bag.

Prepare the material correctly. This of course depends on the type of cutting.

Insert the cutting in a suitable compost. Then make certain the routine work of taking care of it until rooted is never neglected. This consists of regular watering and removal of dead stems or leaves which could form a breeding ground for fungus diseases.

STEM OR TOP CUTTINGS

These offer an easy method of increasing ones stock of many plants which we grow in our gardens. No expensive equipment is necessary, though as interest and experience grow, the range of plants which can be rooted from cuttings will be increased if a heated frame or small propagating unit is installed (1). For small quantities of cuttings, which is usually all the average garden can accommodate, a plant pot enclosed in a polythene bag on the kitchen window sill will often suffice.

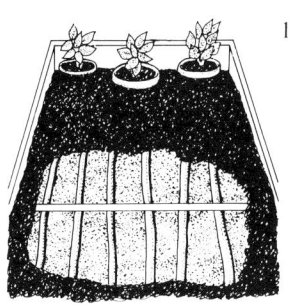

A closed frame, propagating unit or a simple polythene bag provides the close, moist conditions which stop the cutting from drying out until new roots have formed. This is essential with cuttings of soft young shoots taken early in the growing season which de-hydrate very quickly.

The basic ingredient in the compost used for rooting cuttings is sharp, lime free sand. Purchased in small quantities, this may be expensive, but as it can be used over several years with only an occasional topping up to maintain the right depth, it will repay the outlay many times over.

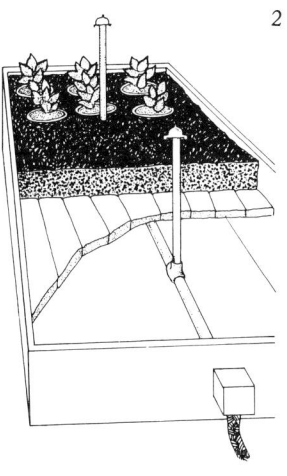

A good, all purpose, compost for rooting cuttings of most plants would be 2 parts of sharp, lime free sand and 1 part of peat. NO fertilizer at all should be included in the mixture as this can damage the roots as they develop.

Good drainage is essential. Roots will not establish in a compost which is sour and airless.

Given the right compost, healthy cutting material and reasonable care it is easy to keep a garden fully stocked with an interesting range of plants.

Refinements such as under-compost heating (1), automatic watering (2) outdoor frames (3) and greenhouses will quicken the process, of course, but are not vital to success.

SOFT WOOD CUTTINGS

These are available early in the growing season, being made from young, soft shoots before they harden. Because they are succulent and full of sap they will quickly dry out. If soft-wood cuttings are not given good conditions after insertion in the compost they will quickly be attacked by fungus diseases which cause damping off. Keep them in a close, fairly humid atmosphere to prevent the leaves losing moisture before roots are formed to replace it. Given favourable treatment, soft wood cuttings should root in 4 or 5 weeks.

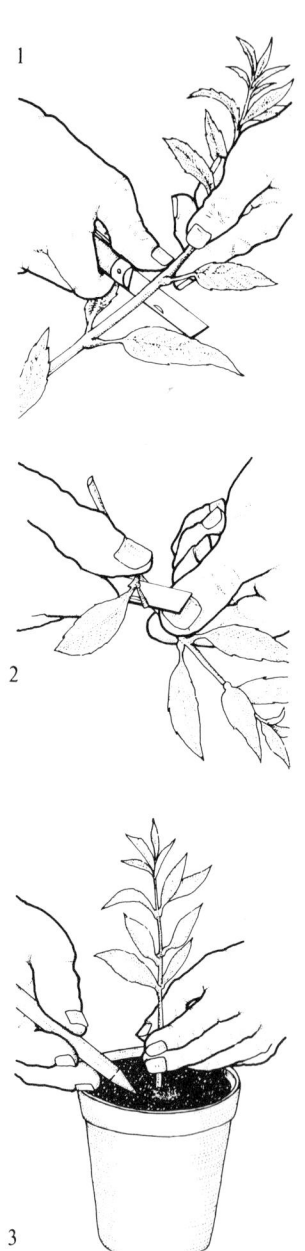

Always use a sharp knife or razor blade when taking cuttings. Make a clean cut just below the node or leaf joint where a leaf joins the stem (1). With one or two exceptions such as clematis or hydrangea, soft wood cuttings are always 'nodal'. The reason is that the concentration of natural hormones within the plant which stimulate rooting are strongest at the junction between leaf and stem. The length of the cutting varies from an inch to four inches, but apart from this the method of propagation is the same. Using a sharp knife or safety razor blade, trim away the lower leaves to above the level at which the base of the cutting will be inserted in the compost (2). Dip the end of the cutting in one of the proprietary rooting powders available at garden shops and insert it in the compost using a blunt ended 'dibber'. A 'dibber' is a word gardeners use to describe a tool used for making holes in the soil. An ordinary lead pencil makes a useful dibber, or one can be made from a piece of hazel or hawthorn smoothed down with a knife.

Push the cutting right to the base of the hole (3). Then firm without crushing the stem and take care that no air pockets are left to gather stagnant air or moisture which will hinder rooting taking place. Cuttings inserted round the edge of a pot form roots more rapidly than those placed in the centre, a point worth remembering with plants which prove troublesome.

When all the cuttings have been inserted in the compost, water them well in. This provides the moist conditions necessary to promote root formation and firms the compost down. Like a seed bed, rooting composts need to be firm.

At first the soft wood cuttings will flag or droop a little, even when shaded as they should be from hot sunshine. Keep them sprayed over with water in a close atmosphere and they will quickly recover. Once roots are formed, the cutting will start to grow. This is a warning that they should be removed, and potted off, into a compost containing a proper food supply. Either John Innes or one of the peat based mixtures will do.

Some herbaceous plants, such as delphiniums, lupins, michaelmas daisies, rot off if they are given too much heat and are best accommodated in an outdoor frame or under glass cloches in a sheltered border.

SEMI-HARDWOOD

These are those prepared from shoots midway through the growing season. Growth is nearing completion but has not ripened to become completely 'woody'. A good check on ripeness is to take a shoot between the fingers and bend it. If the stems snap then the wood is soft, if it is pliable and hardening at the base then it is known as semi-hardwood or half-ripe. This is one of the most widely used methods of increasing many woody plants including heathers, dwarf conifers and small leaved rhododendrons. Half ripe or semi-hardwood cuttings are usually taken from the open garden as opposed to greenhouse plants during the period June to early August.

A mixture of 2 parts sharp, gritty, lime free sand and 1 part peat is, a good general purpose compost for rooting the cuttings into.

Trim the 'heel' or the 'node' to a clean cut with a sharp knife or safety razor blade. Treat the wound with rooting powder after removing the lower leaves, then dibble the cutting into the compost. (This could be the same mixture as that used for soft wood cuttings.) Half ripe cuttings being less succulent than the soft wood cuttings are less prone to wilting, but it is advisable to keep them shaded from direct sunlight for a few days. After that, provided the watering is carried out regularly, warm sunlight will, in most instances, quicken root formation. Semi-hardwood cuttings made and dibbled in during June-July should be well enough rooted to transfer, pot off, into compost containing a proper food supply in October. Newly potted cuttings are best given the protection of a cloche or frame for the first winter.

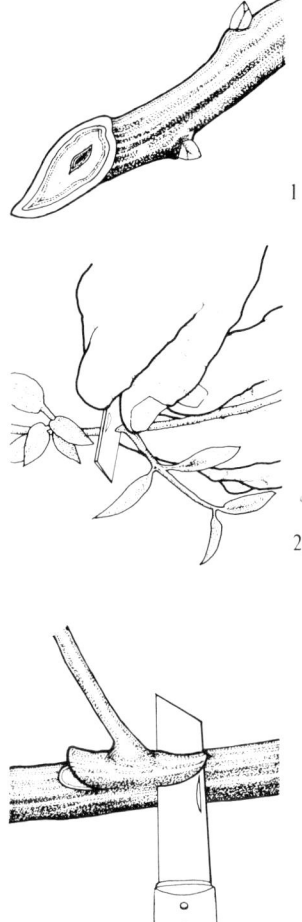

Cuttings from some shrubs will often root better and quicker if the shoot is pulled from the parent branch with a piece of the older wood attached. This is described as a 'heeled' cutting (1).

Sometimes tearing the cutting away with a 'heel' leaves a jagged, unsightly wound on the shrub or tree. I prefer to remove the cutting with a sharp knife though still leaving a piece of the old stem attached. Cut one way (2), then the other (3). This 'heel' is then trimmed to the thinnest wafer possible.

Heeled cuttings are particularly useful with shrubs which have hollow stems which, when taken as normal nodal cuttings, cannot completely callous over so rot off.

Other half ripe cuttings are trimmed to just below a leaf joint and are then referred to as 'nodal' cuttings as described previously.

A few shrubs will root better if the cuttings are taken and trimmed at equal distance between two leaf joints. Clematis are an example. The cutting is then inserted in the compost so the first node is buried about half an inch deep.

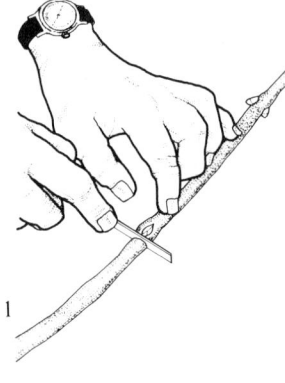

1

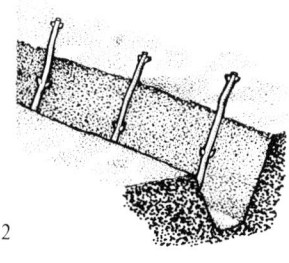

2

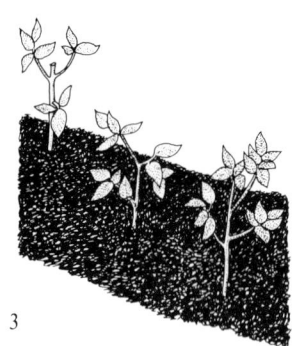

3

HARDWOOD OR BASE CUTTINGS

These are only available late in the growing season as they are prepared from fully ripened wood. The season extends from October to early December with open garden plants and provides a reliable method of propagating roses, gooseberries, blackcurrants and the popular range of garden shrubs like forsythia.

Because the wood is fully ripened, there is not the danger of drying out so they do not respond to the close, humid conditions required by either the soft or semi-ripe shoots. In most cases, attempts to force early growth by keeping hardwood cuttings in close warm conditions do more harm than good. Quite often hardwood cuttings of plants which lose their leaves can be lined out on a sheltered border outdoors and left untended, apart from checking them over at regular intervals to make certain the frost has not loosened them. Cuttings from evergreen shrubs (plants which retain their leaves throughout the winter) can be lined out to make roots in an unheated frame. In each case, cuttings taken in October-December will not be ready for transplanting until the following autumn, approximately twelve months later. So it is fortunate from the gardeners point of view that they require so little attention.

OUTDOOR CUTTINGS

Choose a sheltered border, preferably against a wall and, if the soil is heavy, work in a liberal supply of grit or coarse sand. Take out a straight back trench 2-6 inches deep according to the size of the cuttings to be inserted. Scatter some coarse sand in the trench bottom for the base of the cutting to rest in. I find this helps drainage and stimulates rooting.

It is usual for hardwood cuttings to be 'nodal', so make a clean cut with a sharp knife below a leaf joint (1). Some hollow stemmed shrubs root better if the cutting is taken with a 'heel' (a piece of the parent stem).

The length of cutting varies from 4 to 18 inches, depending on the type of shrub being propagated. Dip the base of the cutting in rooting powder, a stronger type than that used for the soft and semi-hardwood variety. Push the prepared cutting firmly into the sand in the trench, spacing them 3-4 inches apart, vertically against the upright back of the trench (2). Replace the soil and firm it well down round the cuttings with the feet. In a year or two the cuttings should be ready to be transplanted (3).

FRAME

The frame for hardwood cuttings can be prepared by mixing a compost of 1 part loam, 1 part peat and 1 part sand. Loam is really soil which contains plenty of fibre. The best loam is made from grass turf cut from pasture land then stacked for eighteen months to rot before being chopped up and put through a half-inch riddle. The depth of compost required is about 6 inches. The loam is necessary as the cuttings will have to grow for some time in the medium after rooting.

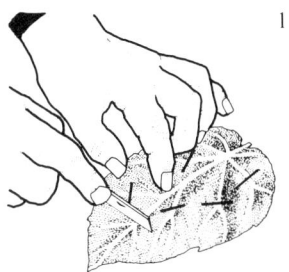

Preparation and insertion of cuttings is the same indoors as out. Trim with a sharp knife, remove the lower leaves, dip in rooting powder then line out in a straight backed trench. The frame will need well watering to settle them in. This task is usually fulfilled more than adequately by the autumn rain. The following autumn the rooted cuttings can be lifted and either lined out for growing on or transferred to a permanent place in the garden.

Rules for taking top cuttings

Make sure the rooting compost is correct with free drainage.
Maintain the most equable temperature possible.
Water the cuttings regularly.
Clean all dead material away before it can serve as a source of infection.
Take cuttings only from healthy, disease free plants.

LEAF CUTTINGS

These are, as the name implies, prepared from healthy leaves just approaching maturity. Not all plants can be propagated by leaf cuttings, but it is a method which has been used successfully, for many years, to increase stocks of plants such as begonia rex, sanseviera, gloxinia, streptocarpus, haberlia, ramonda and saintpaulia.

Begonia Rex Remove a leaf from the parent plant, turn it over on to a hard flat surface; a piece of glass is ideal. Then with a sharp knife make a series of cuts across where the main and secondary veins join (1). Turn the leaf the right way up and peg it flat with some layering pins on a rooting compost of 2 parts sharp sand, 1 part peat (2). Small plantlets should grow from some of the incisions starting at the stalk end. One leaf can be persuaded to produce sometimes up to half a dozen healthy youngsters (3) which are potted off when large enough.

An alternative method is to chop the leaf into 2 inch squares and insert upright into the rooting medium, or lay flat as

previously described. I have not discovered any advantage to this method compared with leaving the leaf whole.

Sanseviera (Mother-in-laws tongue). The leaves should be cut into pieces and pushed into the sandy, peaty compost. These cuttings will only produce mottled green foilage, with no variegation.

Streptocarpus, Gloxinia Push the stalk end of a leaf into the compost just far enough to bury the base of the leaf. A small new plant should grow where leaf and stalk join. Very long leaves can be cut into sections before being dibbled into the compost to root. If the conditions which stimulate rooting are correct, i.e. warmth and humidity, each piece should develop a small plantlet at the base.

LEAF BUD CUTTINGS

These are similar to leaf cuttings but differ in that each leaf is taken with a well developed bud in the axil and a piece of stem also (1). The axil is where the leaf stalk joins the stem. This method is very useful when there is only a small amount of growth from the stock plant and getting top cuttings is impossible. I have taken bud cuttings from camellia by this method at any time from June until February, but I find that the best months are from September to November.

Cut strong, well ripened young shoots into sections each with a leaf and bud attached. With a sharp knife split or scrape the bark on the side furthest away from the leaf to wound it — coat the wounded part in rooting powder, then dibble into compost so the stem is buried, the bud just shows on the surface and the leaf is fully exposed. A compost of 2 parts sand, 1 part peat which must be lime free for camellia, is suitable.

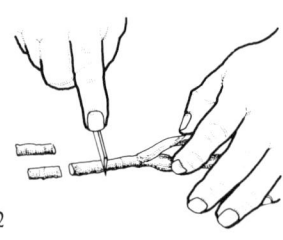

Some plants, notably dracaena and dieffenbachia will root easily from stem sections. I use this method when the plants have lost their lower leaves and begin to look untidy. Cut the stems into inch long sections (2) and either insert them vertically into a compost of 2 parts sand, 1 part peat or lay them horizontally, pressing them into the surface of the mixture until half buried (3). Young shoots with roots attached should grow from these sections and then in 10 to 12 weeks they may be potted off into a peat based, or the John Innes No 1 compost.

Vines are frequently propagated by means of a single bud cut with a short piece of stem. Cuttings like this are referred to as 'eyes'. Lengths of well ripened shoot are removed during the dormant season, then cut into 1½ inch sections, each with a well formed bud in the centre. On the side opposite the bud take off the bark then press into the surface of the compost to root but leaving the bud exposed.

ROOT CUTTINGS

There are plants which have the ability to produce shoots from their roots if these are damaged in any way. Gardeners can take advantage of this by detaching a piece of root, cutting it into 2 to 3 inch long sections and growing new plants from them.

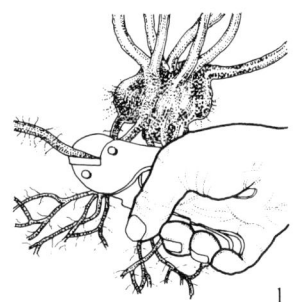

Any plant with a fleshy root system may be tried but by no means all will prove successful. Small specimens can be lifted while the thongs (roots) are selected. Larger herbaceous plants or shrubs can be left in situ only baring the root sufficiently for the required number of thongs to be taken. Cut the roots, those pencil thick will do, with a pair of secateurs or sharp knife close to the main stem (1). Plants such as phlox or primula will rarely have roots pencil thick so I take the most fleshy ones. With a sharp knife, cut the roots into sections 1 to 3 inches long, straight cut at the top sloping at the base.

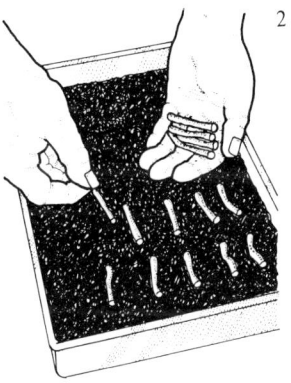

Fill a pot with John Innes seed compost or with one of the peat based mixtures. Then, using a dibber, insert the cuttings flat top up so that it is just covered by the compost. A topping of sharp sand after the pot is filled to about a quarter inch thick will ensure a well drained area for the young shoots to grow through. This lessens the risk of them rotting off. Alternatively the cuttings can be laid in the compost (2) and covered with a quarter inch of sand. This is the system I use for finer rooted plants like the herbaceous phlox and primula.

Water the compost thoroughly. Then stand the pot in a greenhouse frame or protect with a cloche. Shoot and root formation takes about six months, but they are much easier to look after than top cuttings, for there is less risk of them being lost through drying out.

Root cuttings may be taken at any time from September to mid-May.

SCALES

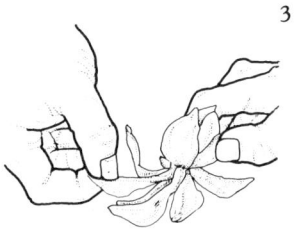

Lilies may be propagated by means of seed, bulbils which form in leaf axils of some species and simple bulb division. Or you can increase stock by means of scales. Scales are the fleshy overlapping leaves which go to make up the bulb. Selected bulbs may be lifted during the dormant season and some of the scales removed. Do not remove too many or the bulb will be weakened. Put the scales into a polythene bag with peat and sand. Shake the contents, then hang them up in a warm place,

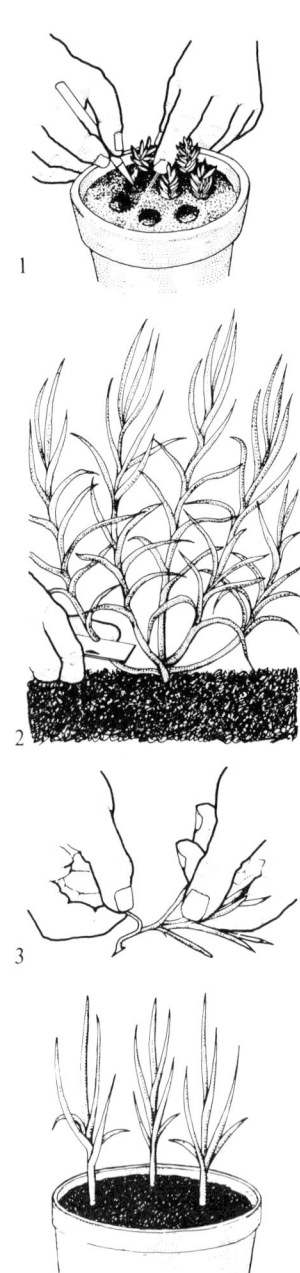

like an airing cupboard, to root. The plants will be ready for potting up individually in 8-10 weeks. They will also root if just planted in trays filled with the same peat and sand mixture.

PLANTS WITH SPECIAL NEEDS

Heather cuttings are made from the short young shoots which grow during the spring and early summer. These can be removed in July or August when they may be anything from three quarters to one and a half inches long. Dip the cuttings in rooting powder and dibble them close together in a compost of 1 peat, 1 sand (1). They should be ready for potting off in 3 to 4 months, but are better left until the following spring when they will have a larger root system.

Some plants have leaves which are covered with hairs. These tend to hold the moisture and so will need special attention. Too much moisture will cause rotting but a dusting with the fungicide Dinocap will help prevent the disease gaining entry. Use a sharp sand as the rooting medium.

Pipings

This is a method of taking cuttings peculiar to pinks and carnations. Young shoots are prepared by simply taking hold of the tip and pulling it or cutting (2) away from the parent plant. The cutting pulls away in very much the same way as a telescope comes to pieces (3), and can be dibbled into a sand compost without any further treatment. Pumice sand is useful for rooting pinks, carnations and many other grey leaved plants.

Aster If continually propagated by division, asters are liable to become infected by eel worm or the fungus disease Verticillium. To make sure shoots to take as cuttings are available early enough to provide flowering plants by the following October, lift strong roots in the autumn and over-winter them in a frame. Stock plants should only be used if they are healthy, free from pest and disease. Alternatively, where space is limited, leave the plants in the ground but cover them with a cloche. Young shoots about 3 inches long will be ready for making into cuttings by mid-March and dibbled into sand will root in 3 weeks.

Aubretia is more often raised from seed, but frequently to propagate a particularly good colour form you must take cuttings. Wait until the plants have finished flowering, usually in late May, then with a sharp pair of shears cut all the top growth back to soil level. To encourage strong young shoots to break after this very drastic pruning I work a little compost over the cut stems. When the young growths are long enough cut them off with a sharp knife and dibble them into sand to root. Plants of viola and pansy can be treated in the same way.

Clematis Clematis will root from cuttings but not with unfailing regularity, some varieties being more predictable than others. Take the semi-ripe cuttings when they are firm for about half their length, making the cut just *above* a leaf joint. Insert the cuttings in a pot filled with sand-peat compost or a frame so that the first node is just buried in the mixture. Water the cuttings well in but afterwards do not let the atmosphere become too close and humid or they tend to damp off.

1

Delphiniums Established plants grow so many shoots in spring that some thinning is necessary or the flower spikes will be small. Surplus growths removed with a sharp knife when they are three or four inches long can be persuaded to root in a frame. No heat is required. In fact, my experience has been that delphinium cuttings in a heated frame are more liable to die off than those rooted cold.

Dahlia Tubers lifted in autumn are stored, after cleaning, in air dry peat to prevent shrivelling. To start the tubers into growth just syringe the peat over with tepid water every day. When the young shoots are about three inches long, remove them with a sharp knife close to the tubers. Dibbled into a sand compost with bottom heat they will root very quickly.

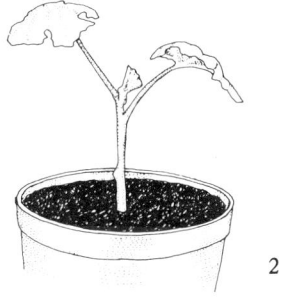

2

Pelargoniums Though comparatively easy to propagate, pelargoniums deserve a special mention, apart from being listed under soft wood cuttings, because they are so often needed in large numbers for summer bedding. Stock plants are selected when the beds are being cleared in the autumn. Label each one so the different colours can be kept separate. I shake the roots clear of soil, then leave them on the bench overnight to wilt. This loss of moisture seems to make the cuttings root better. Choose short jointed shoots 3-4 inches long taken just below a leaf joint (1) and push them into a sandy compost (2). A high temperature is not necessary — a window sill will do; providing the temperature does not fall below 50 degrees F. the cutting will root. Pot them off in March the following year.

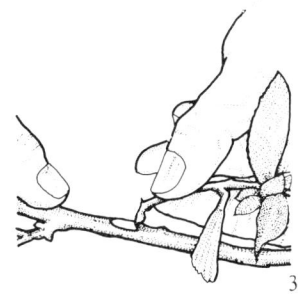

3

Phlox Though perennial, border phlox are usually propagated from root cuttings because the risk of eel worm infection is so great when the plants are just divided. The rock garden or alpine phlox are increased by means of young shoots pulled off with a heel (3) or cut below a node, usually in June. These, when pulled off, root quickly enough in a sand compost.

ALPHABETICAL LIST

Soft wood cuttings

Achillea Taken in early summer

Aethionema By young shoots pulled off in June. Sandy compost

Andromeda Taken in June by tip cuttings

Antirrhinum In July — side shoots or tips

Arabis Taken in June by young shoots

Armeria Taken in June by young shoots

Arnebia By young shoots in June

Aster (Michaelmas Daisy) Taken in March by basal shoots

Aubrieta Taken in July by young shoots

Begonia Taken in March by stem cuttings

Cactus By wilted cuttings May-July

Calceolaria Taken in July by soft wood cuttings

Campanula (alpine species) Taken in May-June by young shoots

Caryopteris Taken in June by soft wood cuttings

Ceratostigma Taken in June by young growth

Cheiranthus (Alpine Wallflower) Side shoots in May-June

Chelone Soft cuttings May-June

Choisya Taken in June by soft wood

Chrysanthemum Taken in December onwards when available from stools

Clematis Taken in May-June

Cobaea Taken in July by cuttings from soft side growths

Coleus Taken in March by cuttings from soft side growths

Coreopsis In June-July by young shoots

Dahlia Taken in February-April by young growths

Delphinium Taken in the spring by young growths

Dianthus Taken in the summer by pipings of non-flowering shoots

Forsythia Cuttings will root at almost any time beginning in June.

Fuchsia Taken in May-June by young shoots

Gaillardia In July — when stock is needed of a specially good colour form.

Gypsophila Taken in June-July by soft wood cuttings

Heliotropium Taken in the spring by soft wood cuttings

Hydrangea Taken in the spring by soft wood cuttings

Impatiens Taken at any time during the growing season by soft wood cuttings

Linum Taken in June as soft wood cuttings

Lobelia Taken in May-June as soft wood cuttings

Lupins Taken in spring as basal cuttings

Mimulus Young shoots as growth starts in spring

Pelargonium (Geranium) Taken in late summer as stem cuttings

Penstemon   Taken in summer as side cuttings
Verbena   Taken in early summer
Viola   Taken in the summer

Semi-hardwood cuttings

Abelia   Taken in July
Abutilon   Taken in July-August
Acacia   Taken in July-August
Alyssum   Taken in May-June
Artemisia   Taken in August, with a heel of old wood
Azalea   Taken in July-August
Berberis   Taken in August
Buxus   Taken in July-August
Camellia   Taken in July, outdoors; in May for pot grown plants
Ceanothus   Taken in July-August
Chaenomeles   Taken in early summer as half ripe with a heel
   of old wood
Cistus   Taken in July-August
Conifers   Taken in June-September
Cornus   Taken in June-July
Coronilla   Taken in June-July
Cotoneaster   Taken in June-August
Cotinus   Taken in June-July
Cytisus   Taken in July-August
Deutzia   Taken in June-August
Diervilla   Taken in July-August
Erica   Taken in June-August
Escallonia   Taken in July-August
Euonymus (evergreen forms) Taken in July-August
Garrya   Taken in August-September
Genista   Taken in July-August as heel cuttings
Grevillea   Taken in July
Hedera (Ivy) Taken in September as cuttings 6-8 inches long
Helianthemum   Taken in June-July
Hibiscus   Taken in July-August
Hypericum   Taken in June-July
Iberis (Evergreen Candytuft) Taken in July-August
Lavendula   Taken in July-August
Lithospermum   Taken in July
Lonicera (Honeysuckle) Taken in July-August with a heel
   of old wood
Magnolia stellata   Taken in June-July
Olearia   Taken in August-September
Osmanthus   Taken in August
Penstemon   Taken in August
Pittosporum   Taken in June-August
Potentilla   Taken in July-August

Pyracantha Taken in July-August
Rhus Taken in July-August
Rosa (Bedding & Shrub) Taken in July-August. Use the shoots which have flowered
Rosmarinus Taken in July-August
Salvia (sage) Taken in July-August
Santolina Taken in July-August
Senecio Taken in August-September
Spiraea Taken in July-August
Syringa (Lilac) Taken in July-August

Hardwood cuttings

Arbutus Taken in November-December but they need a heated frame to root
Buddleia Taken in October-November
Chamaecyparis Taken in September-October cold frame
Cornus Taken in October-November
Deutzia Taken in October-November cold frame
Diervilla Taken in September-October cold frame
Elaeagnus Taken in September cold frame
Forsythia Taken in October
Ilex (Holly) Taken in September-October
Ligustrum Taken in October
Lonicera nitida Taken in October-March
Morus nigra Taken in October
Philadelphus Taken in October-November
Populus (poplar) Taken in October-November
Prunus (laurel) Taken in October
Ribes Taken in October-November
Rosa (roses) Taken in October-November
Salix (willow) Taken in October-March
Skimmia Taken in October-November into a cold frame
Spiraea menziesii Taken in September-October (suckers are usually easier)
Symphoricarpos Taken in October-November
Weigela Taken in October

Root cuttings

Althaea Taken November
Anchusa Taken in October-November
Anemone (japonica) Taken in November
Arnebia echioides October-November
Clerodendrum Taken in March-April
Echinops Taken in Autumn
Eryngium (Seaholly) Taken in November-March
Gaillardia October-November
Papaver (oriental poppy) Taken in November-March

Phlox Taken in October-November
Primula (denticulata) Taken in February-March
Rhus typhina Taken in March-April
Romneya (Tree poppy) Taken in March-May)
Statice (perennial) Taken in October
Yucca Taken in January-March

Leaf cuttings

Begonia rex Taken in summer
Camellia Taken in June-February with a leaf with a bud
  attached
Gloxinia Taken in spring-summer
Haberlea Taken in June-July
Ramonda Taken in June-July
Saintpaulia (African Violet) Taken in spring-summer
Sansevieria (Mother in laws tongue) Taken in May-August
Sinningia (Gloxinia) Taken in April-May
Streptocarpus Taken in March-April
Zygocactus (Christmas Cactus) Taken as leaf sections with three
  or four joints in summer

Layering

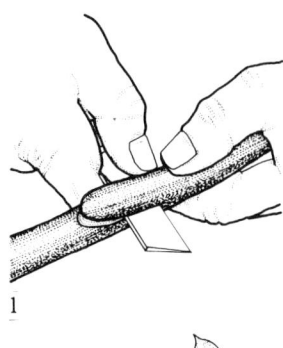

Layering is the term used to describe the easiest and most foolproof method of vegetative propagation. A branch or shoot is persuaded to form roots before it is detached from the parent plant. Once the layer has formed a large enough root system it can be cut away and planted up to lead an independent existence.

BRANCH LAYERING

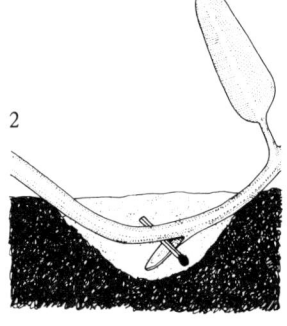

Young, non-flowering shoots which can easily be bent to the ground are the best to choose for rooting from branch layers. With either evergreens or deciduous plants, the season I find most convenient for pulling down the layers is early spring. Growth is just beginning and the roots should form quickly. But there is no set season — layers can be put down at any time.

Prepare the soil under the branch to be layered by forking in a dressing of coarse sand and peat. Pull down the branch to be layered and where the underside touches the soil make a cut to form a tongue 1-2 inches long (cut towards the tip of the shoots) (1). Place a matchstick or piece of sphagnum moss into the slit to prevent the wound closing (2) and healing over instead of rooting. The wounded area is then buried 3 inches deep in the compost and pegged firm with a layering pin made of galvanised wire or wooden peg, leave the growing tip exposed. For extra security this shoot can be tied to a cane (3). A stone on top of the layered branch will help to hold it firmly in position and act as a mulch to conserve moisture. In very dry weather keep the layer well watered and in twelve months it should have developed a good root system. Scrape the soil away in the vicinity of the layer to check on root development before severing the branch from the parent shrub.

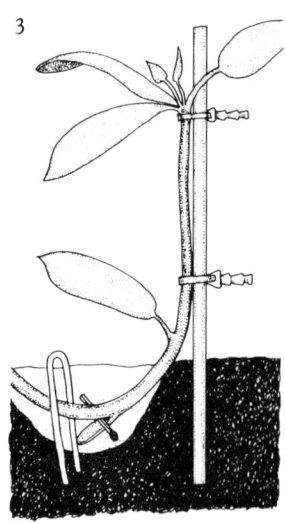

Do not lift the layer immediately you cut it away. Leave the roots undisturbed until the autumn when the transplanting can be done in the assurance that the new plant is capable of surviving alone.

Older branches can be layered successfully but root development is in most cases slower. When planting expensive shrubs like rhododendrons, even though extra stock may not be required immediately, it is sound policy to peg down any handily placed branches as layers.

Herbaceous and soft wooded plants, as for example carnations, should root much quicker — within 3 months compared to the 12 months a hardwood shrub will take to establish. Carnations should be layered in July and roots should have developed well enough for lifting by late September.

TIP LAYERING

This differs from branch layering because the tip of the young shoot only is buried. Loganberries and blackberries are frequently propagated by this method.

Select only young, vigorous shoots which have grown during the spring and early summer. Because the wood is pliable they can be easily bent over and the tips buried 3 inches deep in moist soil (1).

Prepare the soil in advance by forking in a dressing of rotted compost or peat and sand. You can use a 5 inch pot filled with equal parts loam, peat and sand buried near the parent plant. This way the layer will be already potted up when removing the new plants. The shoot tip is then pushed into the mixture. Whichever method is used make sure the tip layer is pegged firmly in position. Rooting should be well enough advanced for the layer to be severed from the parent and transplanted the following spring (2).

RUNNERS

This is a method of self propagation peculiar to some plants. All the gardener needs to do is help out a little by supplying a good compost for the young plantlets to root into. Strawberries are a good example of self layering outdoor plants. Indoors, chlorophytum shows the same alacrity to produce young plantlets at the end of long trailing stems which, if pegged down into good compost, soon root. Violets also increase themselves in this way. A healthy runner will usually give a succession of small plants at intervals, but unless a large stock needs to be built up from limited resources only the first runner need be taken as these tend to grow into the strongest, sturdiest plants.

Select only healthy plants, free from pests or disease to take runners from. Any runners which are not needed should be removed as they begin to grow. The young plantlets may be pegged down into the soil direct so long as this has been enriched with compost or peat. When only a limited number of new plants are needed, for ease of handling the runners can be pegged into pots filled with compost (3) and plunged into the ground at intervals. When rooted, the young plants in pots can be transferred with the minimum root disturbance.

Three inch pots are large enough. Fill with loam or peat based compost. Place the plantlet in position on the compost, still leaving it attached to the parent plant and peg it down with a hooped pin made of galvanised wire or wood. Keep the pots watered in dry weather then by August or September the runners will be well enough rooted to be transplanted.

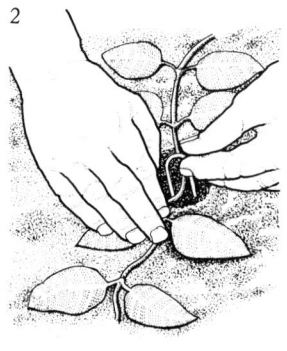

SERPENTINE LAYERING

As the name implies, serpentine layering is used mainly as a means of propagating plants which produce long shoots. These are woven in and out of a well prepared compost, each time burying a leaf joint. Each buried section of the stem must be held firmly in position with a stone or pin. When roots are formed each section is cut away but allowed to remain in position for several weeks before being transplanted. I have found that lifting the rooted layers immediately they are separated from the parent gives a harder check to growth.

Bend the selected shoot down. Where it reaches the soil make a cut in the stem (1) and then peg it into the compost (2). Leave at least two pairs of leaves exposed. Then wound the stem again and bury the damaged area in compost. This is repeated until the whole branch has been treated (3). Root formation should take around 8-12 months. Suitable plants for serpentine layering are jasmine, rambling roses, clematis and honeysuckle.

Serpent layering is usually carried out during June-July. To encourage the production of strong growth suitable for layering older shoots should be cut hard back to the base in spring.

Border carnations are propagated easily by layering during July and August. Choose strong, healthy, disease free shoots which have not flowered. Remove the leaves from the section of stem that will be buried in the compost. Make a cut in the shoot to form a tongue then put in a small splint to hold it open. Bend the shoot down and peg the cut section firmly to the ground. Then cover it with a compost made up of 2 parts loam, 1 part sand, 1 part peat to a depth of 2 inches or so.

Kept well watered, the layers should have rooted in about 8 weeks.

AIR LAYERING

This is a method of propogation which has been made much easier since the introduction of polythene. There are shrubs with branches so brittle or erect that it is almost impossible to pull them low enough to layer at soil level. There are also shrubs not easy to root by cuttings which will quite easily produce roots when air layered. A good example is Ficus elastica (Rubber Plant).

Young straight shoots are the easiest to deal with and root most readily. About half way along the shoot, or about 8 inches from the tip depending on the vigour of the plant being treated, strip off the leaves and make a shallow cut in the centre of the shoot. The cut should be 1½-2 inches in length and about halfway through the stem (1). Dust the wound with rooting powder and put in a piece of match stalk or whisp of sphagnum moss to hold it just slightly apart (2). This is to prevent the cut healing together again instead of growing roots. A sleeve or a sheet made from polythene is wrapped around the branch and tied below the cut with adhesive tape or string. Fill the tube with moist sphagnum moss or a compost made up of sphagnum moss peat and sand (3), then seal the top to prevent moisture loss. This operation should be carried out in May or June, when the plant is in healthy growth.

As with other forms of layering, the risk of failure is reduced to a minimum because root development takes place while the shoot is still attached to the parent plant.

Either coloured or transparent polythene can be used as the plastic sleeve. The see-through type has one advantage, you can watch the roots growing into the mixture. About 10-12 weeks later when the roots have formed, cut the branch from the parent bush below the roots and pot it into a suitable sized pot filled with a sandy compost of 2 parts loam, 1 peat, 1 sand, or the standard potting compost. I prefer a compost without fertilizer for this first potting as the young roots are quite soft and easily damaged. Keep the newly potted plant in a frame or greenhouse, where the air is moist and humid for a week to ten days, increasing the ventilation gradually until it is in a hard enough condition to be transferred to a permanent position.

Old plants of heather, either erica or calluna can also be rejuvenated rather than just digging them out to be burnt. In spring the old plants are lifted, then after the soil has been dressed with peat and sand re-planted so that just the growths at the tips of the old stems are exposed above soil level. In twelve to eighteen months these will have rooted and can be lifted for re-planting in the usual way.

STOOLS

Some shrubs have upright, brittle branches, which when hard pruned will grow a crop of healthy young shoots. If these are earthed up with some sandy soil (4) as they begin to grow, quite often roots develop. The following year, scrape the soil away and sever the rooted shoots close to the base. Cornus, apple or plum root stocks, lilac, azalea, blackcurrants and other shrubs will root when stool layered in this way. Suckers, though possibly a form of plant self layering, fit most readily into division in the next chapter.

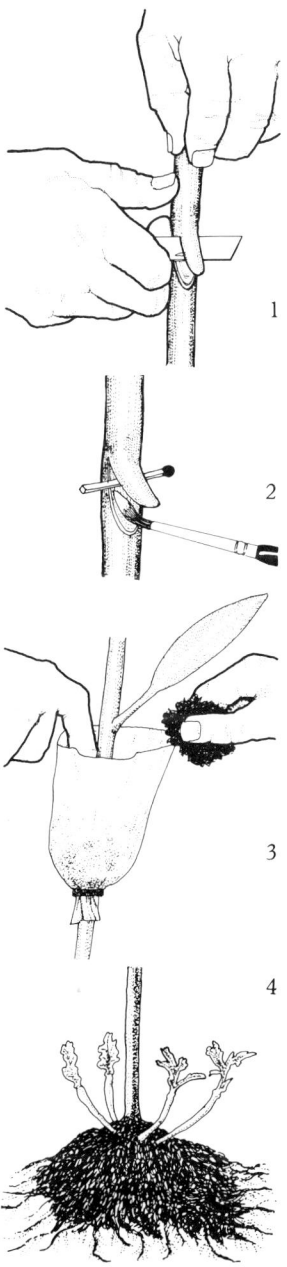

1

2

3

4

ALPHABETICAL LIST

Many shrubs, both indoors and outdoors, may be propagated by means of layering. This method is particularly suitable for house plants which grow too big, for example Ficus (India Rubber plant) The correct time for layering is not exact. I pull layers down at any time during the gardening season though they may take longer to root in the colder months.

Arbutus   Layered in April-May
Aucuba   Ground layer. August-September
Azalea   Layered in April-May
Berberis   Layered in April-May
Camellia   Layered in April-May
Chimonanthus   Layered in April-May
Clematis   Layered in June
Clianthus   Ground layer. June-July
Cornus   Layered in June
Deutzia   Ground layer. August-September
Dianthus (border carnations)   Layered in July
Erica   Layered in April-May
Ficus elastica   Air layering. Summer
Forsythia suspensa   Ground layer. August-September
Lapageria rosea   Serpent layering April-May
Lonicera (honeysuckle)   Layered in June-July
Magnolia   Layered in March
Paeonia   Tree Layered in March
Pernettya   Layered in March
Rhododendron   Layered in March-September
Roses (rambler)   Layered in March-April
Rubus   Layered in June-July
Salix x chrysocoma   Ground layer. August-September
Strawberry   Layered in June
Viburnum   Layered in May-June
Vitis   Ground layer. August-September
Wisteria sinensis   Layered in May

Division

Division is a method of propagation used for any plant which can be broken into several pieces either in situ or after lifting. Each division when planted out will grow into a new plant. Most of the species we grow in the herbaceous border are treated in this way with satisfactory results.

Over the period of four or five years a pyrethrum, michaelmas daisy or shasta daisy will grow into a tangled mass of shoots and roots. One root lifted divided (1) and trimmed (2) will give enough strong healthy shoots (3) with roots attached to begin a new colony elsewhere. Some herbaceous plants, for example paeony and delphinium need very careful treatment if they are to be propagated by division, but the majority present no problems.

Timing of the operation is important, but in general the springtime, just as growth commences, is the safest. Exceptions are primula, pyrethrum and snowdrop which are best dealt with as flowering ends. They seem to re-establish quicker at this time. Obviously, plants with a single stem and root system cannot be increased by division. Anyone just starting a garden can in two years, by exercising a small amount of skill and patience, raise from one or two stock plants a dozen or more healthy divisions. This reduces the cost of buying all the materials needed to fill the garden appreciably.

A brief study of any well stocked herbaceous border will show just how diverse the plants are in root system, habit of growth and time of flowering. The orange red spurge euphorbia 'Fireglow' form a loose tangled mass of roots, each with a shoot which can easily be separated by using a garden fork as a lever (4). Do not be over greedy when dividing overgrown plants, particularly those like sidalcea and scabiosa which form a hard central core. Take only the strong outer portion and divide these into pieces each with a strong root and tuft of shoots attached. Replant immediately before they dry out or wilt. The older, hard, central portion is discarded.

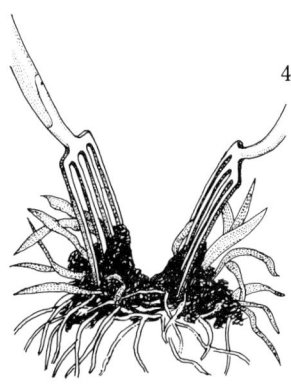

Precautions may be needed when dealing with choice or temperamental plants. You can mix a special compost of peat, loam, sand, plus a dusting of superphosphate to help root growth. Divisions bedded into this mixture establish very quickly. The final resort is to pot up the divisions or bed them out in a frame to overwinter if the specimens are at all tender.

Some herbaceous plants, like the bearded iris, form a fleshy root-like stem at or just below the soil surface. Once again the young roots, each with a tuft or 'flag' of leaves attached, are selected for re-planting. Cut the short, thick rhizome with a sharp knife, dust the cut with sulphur powder, then re-plant so the rhizome just sits on the soil surface.

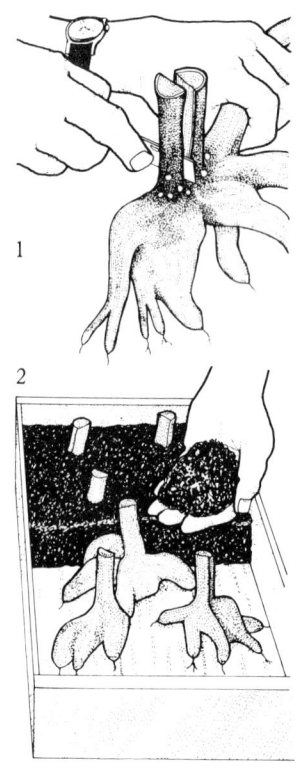

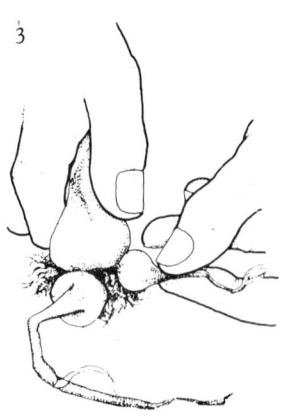

Plants which die back to a swollen underground stem with a cluster of buds on top such as the delphinium cannot be torn apart like their fibrous rooted counterparts, or there is a grave danger that after re-planting the fleshy root will just rot. Make the job a surgical operation. Using a sharp knife cut away one of the buds with a generous section of healthy root attached. This can be bedded in a frame or potted up until growth starts in spring before being replanted out in the garden.

Dahlia tubers require almost identical treatment, though in this case a tuber is formed as an underground foodstore. Lift the plant, then with a sharp knife cut away each tuber which has a piece of stem attached (1). To be really safe, do not divide until spring as growth breaks, then it is easy to see which tubers have the strongest shoots growing from them and a better selection can be made. Bed the divisions in potting compost until well into growth (2). Paeony respond best to this type of treatment also.

BULBS & CORMS

Bulbs and corms are easily propagated because new plants form close alongside the parent every year. In the case of crocus, tulips, snowdrops or narcissus. all you have to do is lift the cluster of bulbs when it is overgrown, break up the mass into single bulbs (3) and replant.

Snowdrops re-establish quicker if lifted and replanted while the foliage is still green just as the flowers fade. With crocus, tulips, and narcissus wait until the foliage dies down.

Gladiolus form a new corm, then as a bonus for the gardener a mass of tiny cormlets or spawn, ranging in size from a cabbage seed to that of a pea. If grown on, this will reach flowering size in a year or two, whereas the young corm formed to replace the parent will flower the following year.

Offsets from bulbs and corms will grow on to display flowers of exactly the same shape and colour with few exceptions.

SUCKERS

Suckers fall mid-way between self layering and division. Because to many gardeners lifting a rooted sucker for transplanting is more akin to division, I have included this method of propagation under the heading division.

Several popular shrubs spread themselves by suckering, for example bamboo, rhus typhina, raspberry, viburnum, romneya etc. This form of self help works to the gardener's advantage, for all that is necessary when new plants are required is to dig up a sucker with a good root, cut it from the parent (1), then replant in the space provided elsewhere in the garden.

Autumn or spring are the best times for transplanting suckers. Though if looked after properly I have moved them successfully in mid-summer. In all forms of division do not attempt to take young plants for propagation from stock which is weak or diseased unless there is absolutely no alternative and then, only if with care, the condition can be cured.

A sucker which comes from below the point of grafting or budding is a product of the rootstock, not the scion. Usually it has no ornamental value. Roses are a common example of plants which are worked on to a rootstock producing suckers which are of no use. Suckers can be removed, planted it rows across the vegetable garden then budded the following year. Rhododendron hybrids are also frequently grafted on the Rh. ponticum rootstock which is very vigorous. Being identical in leaf to the variety worked on to it, a sucker is in many cases not detected until it shows one of the characteristic mauve flowers. When removing suckers check that it is of ornamental value before using it.

There are some plants which if the roots are cut with a knife, spade, or hoe can be stimulated to sucker, for example rhus, some roses, anchusa, baptisa, poplar etc. This method could be resorted to in cases of extreme need – usually the plants named sucker quite happily of their own accord.

PLANTS WITH SPECIAL NEEDS

There are important plants which require special treatment when removing the suckers or offsets. When propagating Globe Artichokes take the strong young shoots which grow from the base of the plant with a sharp knife. Cut off the tops of the leaves before potting them up to overwinter in a cold frame. Then plant out in spring.

Several of the more popular house plants spread by growing suckers or offsets at intervals just a short distance from the parent. The widely grown billbergia suckers so freely that unless these are pulled off regularly the pot soon becomes overcrowded. The best time to do this is in August. Knock the plant out of the pot (2), then pull the offsets away (3) and pot them up (4).

The urn plant (Aechmea) and Flaming Sword (Vriesea) are also easily increased by means of the offsets which grow round the parent plant. These should be removed with a sharp knife and re-potted in a lime-free compost during June-July. Hippeastrum offsets are best removed when the parent bulb is being re-potted.

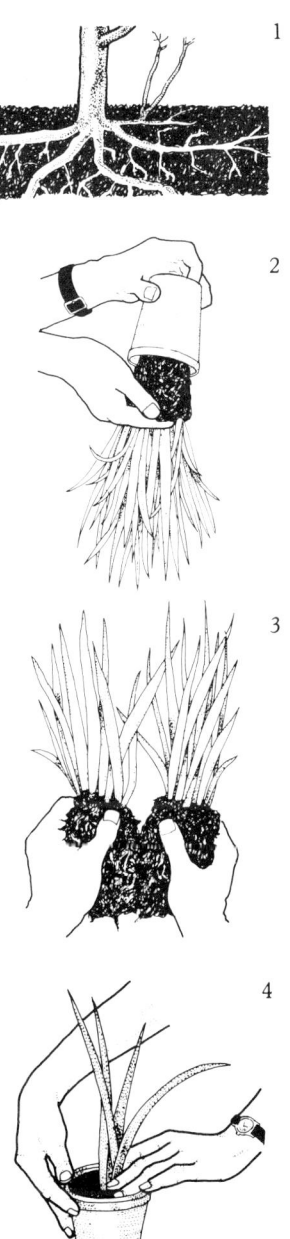

1

2

3

4

ALPHABETICAL LIST

A simple method of propagation when plants form a mass of individual shoots. After division, each piece which has a shoot with a piece of root attached when replanted is capable of growing on into a plant identical with the parent.

Achillea Divide the roots in early spring.

Aconitum Divide the roots in late winter, early spring.

Agapanthus Should be divided just as growth begins in April, though side shoots cut away without lifting the parent plant can be taken at any time.

Allium Divide in autumn and spring.

Alstroemeria Divide the roots in early spring.

Anemone japonica. Divide after flowering in September.

Astilbe Divide the roots in early April.

Bamboo Divide roots just as growth is under way, usually early May.

Baptisia Divide during the dormant winter period.

Bellis Divide immediately after flowering.

Buxus (Box) Divide in spring.

Caltha (Marsh Marigold) Divide in late winter, early spring.

Campanula (herbaceous) Divide in spring.

Catananche Divide in late winter.

Centaurea (herbaceous perennials) Divide in September or early spring.

Chelone obliqua. Divide in early autumn or March.

Chrysanthemum maximum Divide in spring.

Colchicum Divide the bulbs as the foliage dies back in June-July.

Convallaria (Lily of the Valley) Divide the roots in September or March.

Cortaderia (Pampas Grass) Divide in April-May.

Crocus Divide young corms as foliage dies down.

Dicentra Divide in late winter or early spring.

Echinops Divide during the dormant period.

Ferns Divide during March-April.

Galanthus (Snowdrops) Divide just as flowers fade in April

Gentiana Divide in spring, or in the case of G. acaulis after flowering.

Geranium Divide roots in spring.

Geum Divide in late winter, early spring.

Gladiolus Divide corms as foliage dies down. Spawn can also be grown on.

Helenium Divide in spring.

Helianthus Divide in spring.

Helleborus (Christmas & Lenten Rose) Divide roots in March-April.

Hemerocallis Divide roots in spring.

Heuchera Divide roots in March-April.

Hypericum (mat forming species) Divide in spring.

Incarvillea Divide in spring.

Kniphofia (Red-Hot Poker) Divide just as growth begins in spring.

Liatris Divide in April.

Lilium Divide overcrowded bulbs in December-January.

Lychnis Divide in autumn or spring.

Muscaria (Grape Hyacinth) Divide the bulbs in autumn.

Narcissus Divide the bulbs in August.

Nerine Divide the bulbs in July, but only when absolutely essential as they flower better when overcrowded.

Oenothera (perennial forms) Divide in the dormant period.

Paeony Divide the roots as growth starts in spring, making certain that each root section has a healthy young shoot.

Polygonum Divide the roots in March.

Potentilla (herbaceous forms). Divide in March.

Primula (Polyanthus, Primroses etc) Divide immediately after flowering.

Rudbeckia (perennial forms) Divide in spring or autumn.

Salvia (herbaceous forms) Divide in spring.

Saxifraga By offsets or division after flowering.

Scabiosa caucasica. Divide the roots in spring.

Scilla Divide the bulbs in August.

Sedum Divide as growth begins in spring.

Sempervivum Divide when the growth is active in spring.

Sidalcea Divide the roots in spring.

Silene Divide in spring.

Spiraea The majority, even shrubby species, by division in early spring.

Thalictrum Divide as growth commences, replant the pieces in a bed of sharp sand to help them establish.

Thymus (herbal forms) By division.

Trollius Divide the roots in spring.

Veronica Divide during the dormant season.

Budding & Grafting

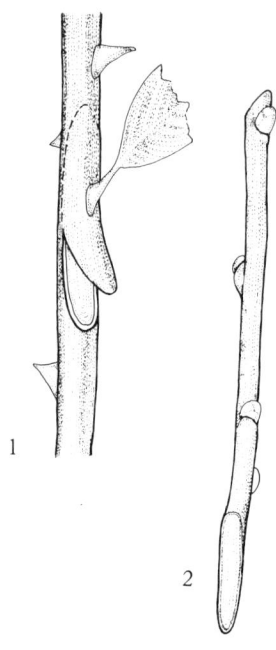

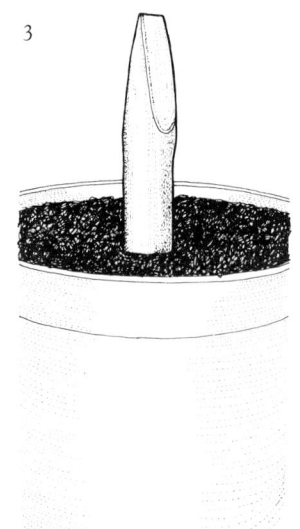

Budding and grafting differ from other methods of propagation in a particular way. A portion of the stem of one individual plant is taken and placed on another entirely different plant in such a manner that if successful they unite and become one. Put in other words the stem or bud of one plant is united with the root of another.

Certain terms are used regularly in describing the technique of budding or grafting.

Bud (1) is precisely what it implies; a single bud with a small piece of bark attached.

Scion (2) is a shoot which is separated from the parent and joined to another in such a way that in maturity it will form the whole branch system or above ground portion.

Stock (3) is the root system of any budded or grafted plant. The prepared stock influences the growth of a plant completely in regard to vigour, ultimate size of tree and the time it takes to reach maturity.

The type of bud or scion used decides what the fruit or flowers will be like. A Cox's Orange Pippin apple grafted onto an East Malling type XXVI rootstock will give a dwarf tree which crops early in life with Cox's Orange Pippin apples. A stronger growing rootstock will do no more than give a larger tree. This will take longer to crop but the fruit will still be Cox.

One of the reasons why plum varieties are difficult to accommodate is that trees on some rootstocks grow very large. By budding onto a less vigorous stock, like St. Julian A, you can get a more manageable bush.

So, though scion and stock are one, each plays an individual part. Scion or bud and rootstock must be compatible, the growing cells have to be capable of uniting. Plants like humans, have a system of tissue rejection. Given care, a certain skill acquired with practice, and the right rootstock for the bud or scion, grafting and or budding are simple enough to perform.

BUDDING

I find this easier than grafting, particularly when applied to roses. Because it is a reliable and economical way of producing rose bushes, as well as many ornamental and fruit trees, it is widely used commercially. Rose stocks can be raised from seed of the wild briar. These are gathered in the autumn, stratified in sand and then sown as described in the chapter devoted to seed sowing. Briars may also be propagated by means of cuttings taken as hardwood in the autumn. Other rose stocks are usually raised from cuttings, for example, laxa, rugosa.

Prepare the soil into which the root stocks are to be planted with care. A plot which has just been cleared of a potato crop is very suitable being well manured, free from weeds and thoroughly cultivated. Plant the stocks in the bed during the autumn for budding in summer the following year. Early November is ideal while the soil is still dry enough to work on. When lining out rootstocks take the same amount of care as when planting an ornamental shrub in the garden. Make certain the hole is dug large enough so the roots can be comfortably spaced out. Nick planting, where the stocks are pushed into a slit made with a spade, is a quick method often used by the nurseryman. Then plant each rootstock by sweeping the roots into the hole (1). The following spring and summer keep the weeds down either by chemical sprays or a push hoe, so that the young stocks can grow strongly and free from competition.

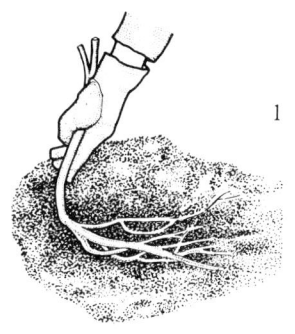

Stocks for standard and half standard roses are treated in the same way but as the stems are 4-6 ft high they may need support. A stage or wire strung between strong posts will prevent them being blown over by gale force winds.

The actual insertion of a bud into a stock can only take place when the bark or 'rind' lifts easily from the wood. This usually happens during the period mid-June to late August, unless the weather is very dry. This reduces the run of sap and the bark tightens onto the wood once more. When this occurs either give the soil a thorough soaking or better, wait until it rains.

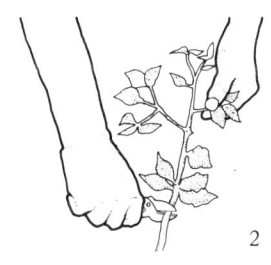

The best buds are usually found on shoots which have flowered and on a stem of at least 12 inches in length (2). Look for a well filled, plump bud. Then as a further test press one or two of the thorns gently sideways (3), they should break easily showing the wood is ripe. Choose the bud sticks from the best example of the variety required with flowers true to colour, of good form and texture and free from disease. Trim away the thorns and leaves, but not the leaf stalk which is left as a handle by which to manipulate the bud. Do not let the bud wood, or the buds, dry out. Wrap them in a moist cloth or stand them in a jar of water.

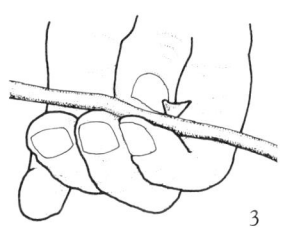

When preparing rose stocks for budding, the soil must be scraped away with a trowel to expose all the stem, as the bud needs to be inserted close above the roots. Another advantage is that the bark, having been buried under moist soil is more elastic and less liable to split when lifted. Some growers even go as far as to draw the soil up around the stock stems in late spring to make certain it is in good condition for budding in summer.

Scrape the soil away (1), then with a damp cloth clean the stem (2) so that no soil is left to get into the wound as the bud is pushed in. The cut to receive the bud is made using a special knife. This has a short, straight, strong blade and a flattened protrusion on the blade or handle end to lever up the bark before pushing in the bud. The cut is made in the shape of a capital T (3). First make the cross cut. Then make another cut to form the stem of the T, upwards to meet the first cut. Using the flattened part of the knife, you can then prise up the two flaps of bark.

Remove the bud from the shoot or 'stick' by making a cut starting an inch below the bud, slanting upwards and coming out the same distance above. Finish with a twisting motion to tear the last strands of bark, leaving a tail (4). Some growers recommend starting above the eye and cutting down to below but this means trimming the base of the shield as a separate process. Which ever way the shield is secured (bud and bark are collectively known as a shield) a piece of wood will be attached. Remove this by easing back the bark (5). Then insert the knife and by twisting the heel of it the bark should fly out. Check to make certain the centre of the bud is still intact as sometimes this flies out with the bark sliver. The embryo of the bud appears as a white dot on the inner shield (6). Always handle the shield by the stump of the leaf stalk. Take care not to touch the inner side of the shield too much, as sweating hands could prevent the bud 'taking'. Lift the bark at the top of the T cut and slide the shield under; an operation made easier because the natural curve of the shield fits to the curve of the stock (7). Trim off the upper end of the shield so it fits snugly under the flaps of the T shaped cut (8). Tie the flaps firmly into position with moist raffia (9) or one of the special rubber or plastic bud ties made specially for the purpose. Do not pull the ties too tight (10) or they constrict the stem. Replace the soil up to the BASE of the grafted bud only. DO NOT bury the bud below the soil level. Normally one graft is done on one root stock.

A good check on whether the bud has taken is that the wounded area turns green. Also the stump of the leaf stalk drops away.

In mid-late February the following year all the top growth above the bud is removed with a sharp pair of secateurs (11). As the shoot from the bud grows tie it in to a short cane or it may get broken off by strong winds. When new growth is a few inches long, pinch it back to 2 buds (12).

Standard roses are budded using the same technique, but the stock is a stem of briar or rugosa 4-6 ft high. On briars

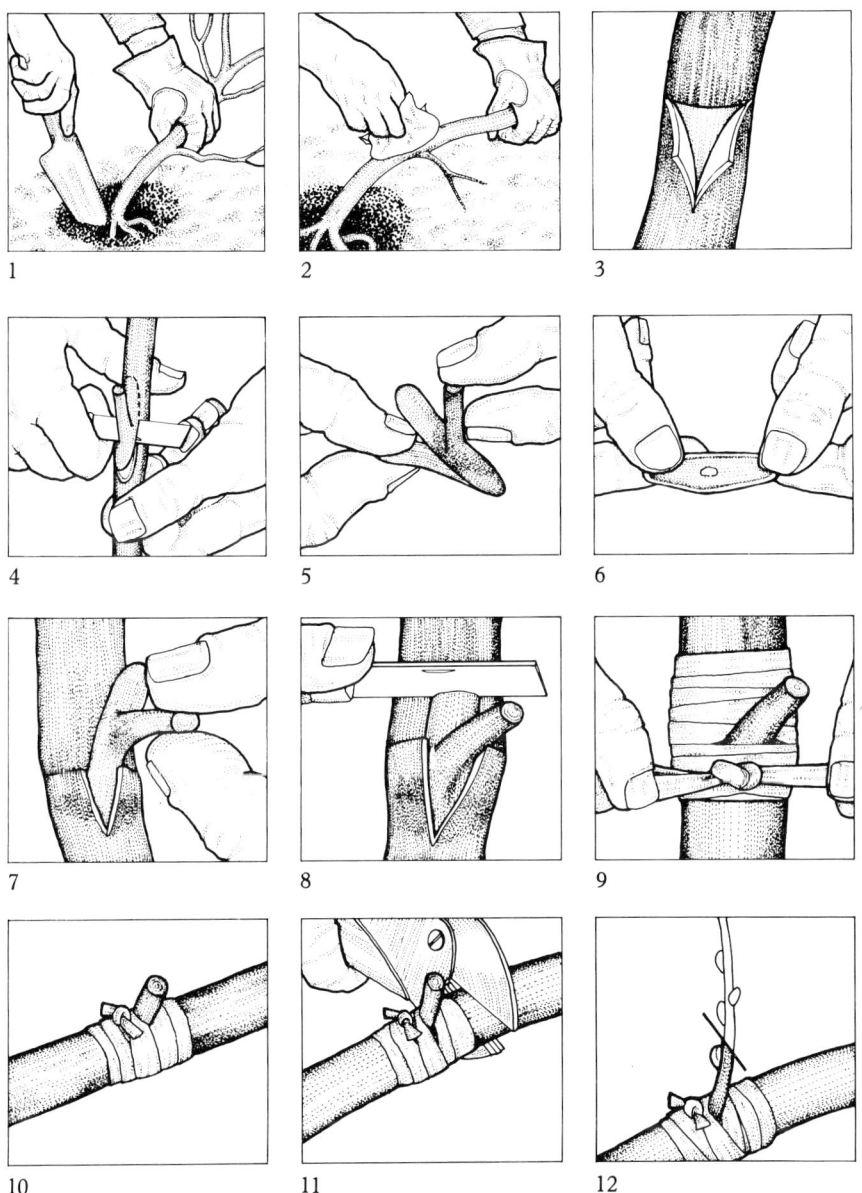

1
2
3
4
5
6
7
8
9
10
11
12

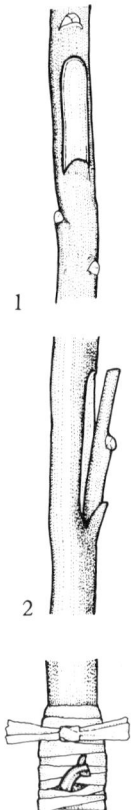

1

2

3

insert the buds on the upper sides of side shoots close to the main stem. Use 3 shoots, 1 bud in each which in due course will grow out to form the head or branch system. Choosing the buds, taking and grafting them into the stock requires the same technique as the one used for bush roses.

Budding of fruit and ornamental trees onto selected root stocks is usually done during July and early August. Rootstocks for budding should be planted up a year previous to being worked.

Quite often children will sow apple pips, cherry stones etc., they germinate then because growth is so slow the interest dies. Growing any sort of hybrid from seed to flower or fruiting is a gamble — the end product could be worthless, but instead of throwing the seedlings away grow them on for use as rootstock. Line them out in the same way as the selected rootstocks, then when the stems are the thickness of a pencil they are ready for working. Unlike the selected rootstocks where the gardener knows what size tree they will produce, seedling rootstock are an unknown quantity, but well worth practising on.

Two methods of budding are used. In one a T shaped cut is made in the stock as described for roses. Buds should be taken from healthy shoots having grown during the current season. Cut off the leaf blade, leaving the stalk protruding to serve as a handle. Slice the bud off with a sharp knife, starting three quarters of one inch below the bud and emerging the same distance above. Insert the bud into the T shaped cut on the stock, trim the top level with the cross cut and tie firmly, so the bark of the stock overlaps and holds the bud shield.

Point of budding depends on the form of tree required. Bush trees will be budded 3-6 inches above soil level; or up to 6 ft for a standard. In a late season the scion wood selected to provide the buds in July should have the soft growing top removed.

CHIP BUDDING

This is a development from T working. A cut is made in the stock peeling a strip of bark some 1½-2 inches long away completely, but leaving a small lip at the base (1). A bud is then cut from a young shoot of the current seasons growth of a size which will more or less cover the wound made in the stock (2). The lower end fits under and is held by the basal lip. The whole is then tied firmly in place with raffia, plastic strip or polythene tape. (3). With both T and chip grafts, in February cut away all growth from the original stock above the inserted bud. As growth breaks from the bud tie it securely to a cane to prevent damage by the March winds.

Grafting

Unlike budding, grafting is usually performed during March or April. To be successful, the stock needs to be well into growth and the scion dormant. The scion is a length of one year old wood 3-5 buds long taken during the winter from a suitable parent plant. To make sure the scions stay dormant until needed for grafting, tie them in bundles and push the ends into moist sand on the north side of a wall or building.

Make sure all the cuts are made cleanly, without tears as this would make uniting the two cuts (graft and scion) more difficult. I find a strong, straight bladed knife the easiest to use and keep this well honed for the purpose.

Wide, soft raffia is the tying material I have always used, though synthetic substitutes are now more readily obtainable.

Stock for grafting needs to be lined out a year prior to being worked so they are well established and capable of vigorous growth. The nearer the stock and scion are to matching in size the better chance there is of the cambium layers being in continuous contact. Cambium is the area of active growth immediately below the bark — present on both stock and scion. A union is not possible if the two cambium fail to touch for at least part of their length.

WHIP AND TONGUE GRAFTING

You prepare the rootstock by cutting it back to 3-6 inches above the soil level. Do this by making a slanting cut upwards some 1½-2 inches long. A small downward cut is then made to form a tongue near the top of the slanting cut (1). Prepare the scion by making a matching cut of the same shape and length. Put a tongue, lipped the opposite way to the one made in the stock, so that they fit and are held neatly together (2). Make sure the surfaces fit flush along their length with the cambium matching on one side at least. Bind the graft with tape or moistened raffia (3), then paint with grafting wax to make it waterproof. Then cut the top off the scion. This wound should also be waterproofed with the wax.

SADDLE GRAFTING

This is another method of propagation used in the nursery trade mainly for rhododendrons. Stock and scion must be more or less the same size. For rhododendrons the rootstocks are potted up and started into growth ready for grafting in February-March. Scions of one year old wood, each with a healthy terminal bud are collected from bushes grown outdoors as required.

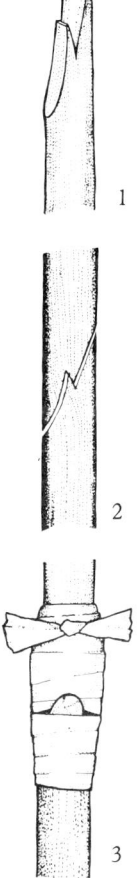

1

2

3

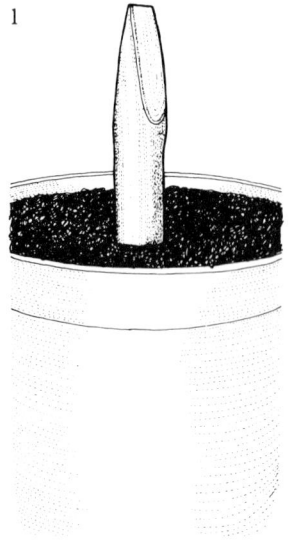

1

Prepare the stock with two oblique cuts meeting in the centre to form a wedge like an inverted V (1). A similar cut is made in the scion so that it sits neatly astride the stock (2). The two are then bound tightly together (3) and the joint made waterproof with grafting wax.

RIND OR CROWN GRAFTING

Old trees, or those worked with a poor quality variety can be often made economicaly worthwhile if they are crown grafted. Cut the old branches off, then insert scions cut into the shape of a wedge at the end into vertical cuts in the bark. Large branches are worked with three scions, smaller branches with two. Bind the scions firmly in place with tape and seal with grafting wax.

After-care of the scions once they start into growth is also important or the tree becomes a tangle of branches, carrying a crop of fruit which is too small. Begin pruning when the scions have made two years growth.

BRIDGE GRAFTING

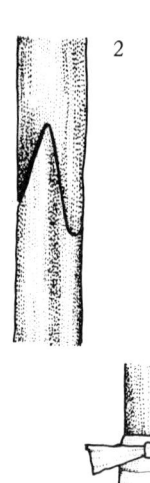

2

I found this method very effective when hares did damage to some five year old apple trees during a month of cold weather. The bark was eaten completely away at about 25 inches above soil level effectively cutting off the food supply to the branches.

The scion is prepared in the same way as for rind grafting, but the wedge shape is made at both ends — not just one. Trim the damaged wood to be bridged so that slit like incisions can be made into healthy bark on opposite sides of the wound. The prepared scions are then pushed into the slits and secured so that they act as channels to conduct sap to the branches. A large branch may need five or six bridges to be effective. When sufficient are inserted, tie them firmly and make all secure with grafting wax. In due course, the bark will grow over to seal the wound completely.

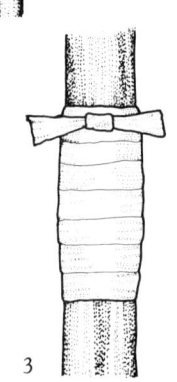

3

The after care of all grafted plants is important and to neglect it can make all the difference between success and failure. Once the scion has made several inches of growth, indicating it has united with the stock, cut the ties very carefully which are holding it in place. DO NOT peel them away — leave them to do that naturally. Sometimes when ties are left they cut through into the scion and restrict its growth. Any buds which break from the rootstock below the point of union with the scion should be rubbed out before they compete or become capable of being confused with scion growth.

ALPHABETICAL LIST

Many ornamental trees which cannot be propagated by seed, layers, or cuttings are grafted.

Acers Named varieties onto rootstocks of species grown from seed.

Aesculus hybrids budded onto seedling horsechestnut, in July-August.

Apple onto Malling or Malling Merton.

Arbutus (Strawberry Tree) Varieties on to a unedo seedling.

Azalea onto Rhododendron luteum.

Betula Varieties on to seedlings of common birch.

Carpinus Varieties on to seedling of common hornbeam. (C.betulus).

Castanea (chestnut) varieties budded July-August.

Clematis onto Clematis vitalba stocks.

Conifers onto seedlings from which the variety has been developed.

Cotoneaster hybrids As standards budded July on to C.frigida seedlings.

Crataegus (Hawthorn) Varieties on to common thorn.

Fagus (Beech) Named varieties are grafted onto seedlings of the common beech Fagus sylvatica.

Fraxinus (Ash) onto seedlings of F.excelsior—Spring.

Hamamelis Are grafted, but layering gives the best results for the amateur. Usually on to seedling H. virginiana.

Ilex (Holly) onto common holly seedlings.

Juglans Varieties On to seedlings of J.regia.

Laburnum Named varieties onto common laburnum seedlings.

Liriodendron (Tulip tree) varieties onto seedlings of the species. March.

Malus (crab apple) onto seedling of Common Crab.

Nectarine onto St. Julian A rootstock.

Peaches onto St. Julian A rootstock.

Pears onto Malling Quince A or C.

Plums onto St. Julian A rootstock.

Prunus (cherries) onto Malling 12/1 or seedlings.

Quercus (oak) varieties onto seedlings of Common Crab.

Rhododendrons Named varieties onto Rh. ponticum rootstock in a heated frame.

Robinia vars onto seedlings of Robinia pseudo-acacia.

Roses Budded usually from late June-August.

Syringa (Lilac) Budded onto common lilac, syringa vulgaria.

Tilia vars (lime) onto layers of Common Lime.

Wisteria Varieties on Wisteria sinensis.

Index

A

Abelia, 31
Abutilon, 31
Acacia, 31
Acer, 51
Achillea, 42
Aconitum, 42
Aechmea, 41
Aesculus, 51
Aethionema, 30
African Violet, 17, 33
Agapanthus, 42
Ageratum, 14
Allium, 42
Alstroemeria, 42
Althea, 32
Alyssum, 14, 31
Amaranthus, 14
Anchusa, 32, 41
Andromeda, 30
Anemone japonica, 32, 42
Antirrhinum, 14, 30
Apple, 51
Apple, rootstocks, 37
Arabis, 30
Arbutus, 32, 38, 51
Armeria, 30
Arnebia, 30
Artemisia, 31
Ash, 51
Aster, 14, 28, 30
Astilbe, 42
Aubretia, 14, 28
Aucuba, 38
Azalea, 31, 37, 38, 51

B

Bamboo, 42
Baptisia, 42
Beech, 51
Begonia, 14, 30
Begonia rex, 25, 33
Bellis, 42
Berberis, 31, 38
Betula, 51
Billbergia, 41
Blackberries, 35

Blackcurrants, 24, 37
Budding, 44
    bud, 44
    chip, 48
    fruit trees, 48
    knife, 46
    roses, bush, 45, 46
    roses, standard, 45, 46
    shield, 46
    trees, ornamental, 48
Buddleia, 32
Bulbils, 27
Bulbs, 40
Butterfly flower, 18
Buxus, 31, 42

C

Cabbage, 19
Cactus, 14, 30
Calceolaria, 14, 30
Calendula, 14
Caltha, 42
Calluna, 13
Camellia, 26, 31, 33, 38
Campanula, 14, 30, 42
Candytuft, 16
Carnations, 15, 28, 34, 38
Caryopteris, 30
Castanea, 51
Catananche, 42
Cauliflower, 19
Ceanothus, 31
Centaurea, 42
Ceratostigma, 30
Chaenomeles, 31
Chamaecyparis, 32
Cheiranthus, 14, 30
Chelone, 30
Chelone, obliqua, 42
Chestnut, 6
Chimonanthus, 38
Chlorophytum, 35
Choisya, 30
Christmas Cactus, 33
Chrysanthemum, 14, 30, 42
Cineraria, 15
Cistus, 31
Clarkia, 15

Clary, 15
Clematis, 22, 23, 30, 36, 38, 51
Clianthus, 38
Clerodendron, 32
Cloches, 11
Cobaea, 30
Coleus, 15, 30
Colchicum, 42
Compost, 6
  lime-free, 7
  mixing, 8
  peat, 7
  seed, 7
Compost, soil based, 7
  peat based, 7
  watering, 7
Conifers, 31, 51
  dwarf, 23
Containers, types 8
  filling, 8
  watering, 10
Convallaria, 42
Convolvulus, 15
Coreopsis, 30
Cornflower, 15
Cornus, 31, 32, 37, 38
Coronilla, 31
Cortaderia, 42
Cosmos, 15
Cotinus, 31
Cotoneaster, 11, 13, 15, 31
Crab apple, 51
Crataegus, 15
Crocus, 40, 42
Cuttings, 20
  base, 23
  half-ripe, 23
  hardwood, 24
  heeled, 23
  how to take, 20
  in a frame, 25
  leaf, 20, 25
  leaf bud, 26
  outdoor, 24
  root, 20, 27
  semi hardwood, 22
  soft wood, 21, 22
  stem, 20, 21

top, 20, 21
  when to take, 21
Cyclamen, 13, 15
Cytisus, 31

D
Dahlia, 29, 40
Daphne, 15
Deutzia, 31, 38
Delphinium, 13, 15, 22, 29, 30, 39, 40
Dianthus, 15, 30, 38
Dibber, 5, 22
Dicentra, 42
Dieffenbachia, 26
Diervilla, 31
Digitalis, 15
Division, 39
Dracena, 26

E
Echinops, 32, 42
Elaeagnus, 32
Erica, 13, 31, 38
Eryngium, 32
Escallonia, 31
Eschscholzia, 15
Euonymus, 15
  evergreen, 31
Euphorbia, 39
Evening Primrose, 17

F
Fagus, 50
Ferns, 42
Ficus, elastica, 36, 38
Forget-me-not, 17
Forsythia, 30, 32
Foxglove, 15,
Frame, 10, 25
Fraximus, 51
Freesia, 15
Fruit trees,
  budding, 48
Fuchsia, 30

G
Gaillardia, 32

Galanthus, 42
Garrya, 31
Geum, 42
Genista, 31
Gentiana, 10, 13, 15, 42
Geranium, 15, 42
Germination, 10
Gladiolus, 40, 42
Globe Flower, 18
Gloxinia, 15, 25, 33
Godetia, 15
Galanthus, 42
Gooseberry, 24
Grafting, 44, 49
  saddle, 49
  whip and tongue, 49
Grape Hyacinth, 43
Grasses, 15
Grevillea, 31
Gypsophila, 15, 30

H
Haberlea, 33
Hamamelis, 15, 51
Hawthorn, 11, 15
Heather, 23, 31
Hedera, 31
Helenium, 42
Heliotrope, 30
Helianthemum, 31
Helianthus, 42
Helleborus, 15, 42
Hemerocallis, 42
Heuchera, 43
Hibiscus, 31
Holly, 11, 13, 32
Hollyhocks, 16
Honesty, 16
Honeysuckle, 31, 34, 38
Hydrangea, 22, 30
Hypericum, 31, 43

I
Iberis, 16, 31
Ilex, 32, 51
Impatiens, 16, 30
Incarvillea, 16, 43

Ipomoea, 16
Iris, 40
Ivy, 31

J
Jasmine, 36
Juglans varieties, 51

K
Kniphofia, 43

L
Laburnum, 16, 51
Lapageria rosea, 38
Larkspur, 16
Lathyrus, 18
Lavatera, 16
Lavendula, 31
Layering, air, 36
  branch, 34
  serpentine, 36
  tip, 35
Leeks, 19
Liatris, 43
Ligustrum, 32
Lilac, 37
Lily (Lilium), 16, 27, 43
  of the valley, 42
Limnanthes, 16
Linum, 16, 30
Liriodendron, 51
Lithospermum, 31
Livingstone Daisy, 16
Loam, 7
Lobelia, 9, 16, 30
Loganberries, 35
Lonicera, 32, 38
Love in a mist, 17
Love lies bleeding, 16
Lunaria, 16
Lupin, 16, 22, 30
Lychnis, 16, 43

M
Magnolia, 6, 13, 16, 38

Malus, 51
Marigold, 16
Marsh marigold, 42
Matricati, 16
Matthiola, 16
Mecanopsis, 10, 13, 16
Mesembryanthemum, 16
Michaelmas daisy, 22, 39
Mignonette, 16
Mimulus (Monkey Musk), 16
Morning Glory, 16
Morus niga, 32
Mother-in-laws tongue, 33
Muscari, 43
Myosotis, 17

N
Narcissus, 40, 43
Nasturtium, 17
Nemesia, 17
Nectarine, 51
Nerine, 43
Nicotiana, 17
Night Scented Stock, 16
Nigella, 17
Node, 21, 22

O
Oak, 6, 51
Oenothera, 17, 43
Olearia, 31
Osmanthus, 31

P
Paeony, 6, 13, 17, 38
Pampas Grass, 42
Pansy, 17
Papaver, 17, 32
Peaches, 51
Peach, stone, 13
Pears, 51
Pelargonium, 29, 30
Pernettya, 38
Penstemon, 16, 31
Petunia, 17

Philadelphus, 32
Phlox, 17, 29, 32
Pieris, 13
Pittosporum, 31
Plums, 51
Polyanthus, 17
Polygonum, 43
Poppy, 17
Populus (poplar), 32
Potentilla, 31, 43
Potting, off, 12
  compost, 7, 8
Pricking out, 12
Primrose, 17
Primula, 13, 17, 33, 39, 43
Prunus, 51
Pyrethrum, 17, 39
Poppy, 32
Pyracantha, 32

Q
Quercus, 51

R
Ramonda, 33
Red Hot Poker, 43
Raspberry, 40
Reseda, 16
Rhododendron, 13, 23, 38, 49, 51
Rhus, 32
Rhus, typhina, 33
Ribes, 32
Riddle, 5
Robinia vars, 51
Romneya, 33
Rooting powder, 22, 25
Rose (Rugosa), 10, 13, 17, 24, 30, 32, 41, 51
  budding, 44, 45, 46
  preparation of stocks, 45
  rambling, 36, 38
Rosmarinus, 32
Rubber Plant, 36
Rubus, 38
Rudbeckia, 17, 43
Runners, 35

INDEX

S
Sage, 32
Saintpaulia, 17, 33
Salad crops, 10
Salix, 32
Salix x chrysocoma, 38
Salpiglossis, 18
Salvia, 18, 32, 43
Sanseviera, 25, 33
Santolina, 32
Saxifraga, 18, 43
Scabiosa, 39, 43
Scales, 20, 27
Scilla, 43
Scion, 44
Seaholly, 33
Sedum, 43
Seed, frame, 10
  hard coated, 13
  outdoors, 11
  pelleted, 10
  protection of, 11
  sowing, 6, 9, 10, 11, 12
  storing, 6
  vegetable, 11
Seedlings, transfer of, 12
Sempervivum, 43
Senecio, 32
Shasta daisy, 39
Sidalcea, 39, 43
Silene, 43
Sinningia, 33
Skimmia, 32
Snowdrop, 39, 40, 42
Solanum capsicastrum, 18
Spawn, 40
Spiraea, 32, 43
Statice, 18
Stock, 44
Stocks, 18
  East Lothian, 18
Stools, 37
Stratification, 13
Strawberries, 35, 38
Streptocarpus, 25, 33
Suckers, 40
Sweet pea, 18
Symphoricarpus, 32

Syringa, 51

T
Tagetes, 18
Thalictrum, 43
Thymus, 43
Tilia vars, 51
Tobacco flower, 17
Tree Poppy, 33
Trollius, 18, 43
Tulip, 40

V
Vegetables, 18
Verbena, 18, 30
Veronica, 43
Viburnum, 38
Viola, 18, 30
Vine, 26
Vitis, 38

W
Wallflower, 18
Weigela, 32
Willow, 32
Winter Cherry, 18
Witch Hazel, 51
Wisteria, 38, 51

Y
Yucca, 33

Z
Zinnia, 18
Zygocactus, 33